# Fuel Production with Heterogeneous Catalysis

# Fuel Production with Heterogeneous Catalysis

Editor

**Manoj Karkare**

**Fuel Production with Heterogeneous Catalysis**

Edited by **Manoj Karkare**

Printed in 2017

ISBN: 978-1-68117-030-5

Library of Congress Control Number: 2015931825

© 2016 by
SCITUS Academics LLC,
616, Corporate Way, Suite 2, 4766,
Valley Cottage, NY 10989

www.scitusacademics.com

# Contents

vi

# Preface

A catalyst is another substance than reactants products added to a reaction system to alter the speed of a chemical reaction approaching a chemical equilibrium. It interacts with the reactants in a cyclic manner promoting perhaps many reactions at the atomic or molecular level, but it is not consumed. Another reason for using a catalyst is that it promote the production of a selected product.

A catalyst that is in a separate phase from the reactants is said to be a heterogeneous, or contact, catalyst. Contact catalysts are materials with the capability of adsorbing molecules of gases or liquids onto their surfaces. The great majority of practical heterogeneous catalysts are solids and the great majority of reactants are gases or liquids. Heterogeneous catalysis is of paramount importance in many areas of the chemical and energy industries. An example of heterogeneous catalysis is the use of finely divided platinum to catalyze the reaction of carbon monoxide with oxygen to form carbon dioxide.

Fuel Production with Heterogeneous Catalysis presents the groundbreaking discoveries, recent developments, and future perspectives of one of the most important areas of renewable energy research—the heterogeneous catalytic production of fuels.

**Editor**

# Chapter 1

# Heterogeneous Catalysis for Sustainable Biodiesel Production via Esterification and Transesterification

Adam F. Lee, James A. Bennett, Jinesh C. Manayil, and
Karen Wilson

European Bioenergy Research Institute, Aston University, Aston
Triangle, Birmingham B4 7ET, UK.

## INTRODUCTION

Sustainability, in essence the development of methodologies to meet the needs of the present without compromising those of future generations, has become a watchword for modern society, with developed and developing nations and multinational corporations promoting international research programmes into sustainable food, energy, materials, and even city planning. In the context of energy, despite significant growth in proven and predicted fossil fuel reserves over the next two decades, notably heavy crude oil, tar sands, deepwater wells, and shale oil and gas, there are great uncertainties in the economics of their exploitation *via* current extraction methodologies, and crucially, an increasing proportion of such carbon resources (estimates vary between 65–80%[1–3]) cannot be burned without breaching the UNFCC targets for a 2 °C increase in mean global temperature relative to the pre-industrial level.[4,5] There is clearly a tightrope to walk between meeting rising energy demands, predicted to climb 50% globally by 2040[6] and the requirement to mitigate current $CO_2$ emissions and hence climate change. Similar considerations apply to ensuring a continued supply of organic materials for applications including polymers, plastics, pharmaceuticals, optoelectronics and pesticides, which underpin modern society, and for which significant future growth is anticipated, tracking the predicted four-fold rise in global GDP and associated requirements for advanced consumer products by 2050.[7] The quest for sustainable resources to meet the demands of a rapidly rising world population represents one of this century's grand challenges.[8,9] Heterogeneous catalysis has a rich history of facilitating energy efficient selective molecular transformations and contributes to 90% of chemical manufacturing processes and to more than 20% of all industrial products.[10,11] In a post-petroleum era, catalysis will be central to overcoming the engineering and scientific barriers to economically feasible routes to alternative source of both energy and chemicals, notably bio-derived and solar-mediated *via* artificial photosynthesis (Scheme 1).

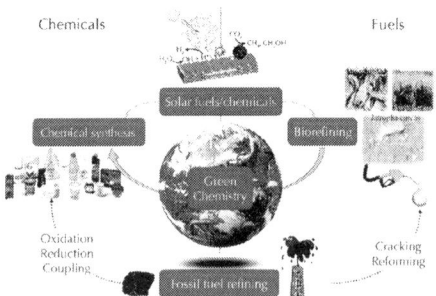

**Scheme 1:** Current and future roles for heterogeneous catalysis in the production of sustainable chemicals and fuels.

While many alternative sources of renewable energy have the potential to meet future demands for stationary power generation, biomass offers the most readily implemented, low cost solution to a drop-in transportation fuel for blending with/replacing conventional diesel[12]*via* the biorefinery concept, illustrated for carbohydrate pyrolysis/hydrodeoxygenation (HDO)[13,14] or lipid transesterification[15,16] to alkanes and biodiesel respectively in Scheme 2. First-generation bio-fuels derived from edible plant materials received much criticism over the attendant competition between land usage for fuel crops *versus* traditional agricultural cultivation.[17] Deforestation practices, notably in Indonesia, wherein vast tracts of rainforest and peat land have been cleared to support palm oil plantations, have also provoked controversy.[18] To be considered sustainable, second generation bio-based fuels and chemicals are sought that use biomass sourced from non-edible components of crops, such as stems, leaves and husks or cellulose from agricultural or forestry waste. Alternative non-food crops such as switchgrass or *Jatropha curcas*,[19] which require minimal cultivation and do not compete with traditional arable land or drive deforestation, are other potential candidate biofuel feedstocks. There is also growing interest in extracting bio-oils from aquatic biomass, which can yield 80–180 times the annual volume of oil per hectare than that obtained from plants.[20] Around 9% of transportation energy needs are predicted to be met *via* liquid biofuels by 2030.[21]

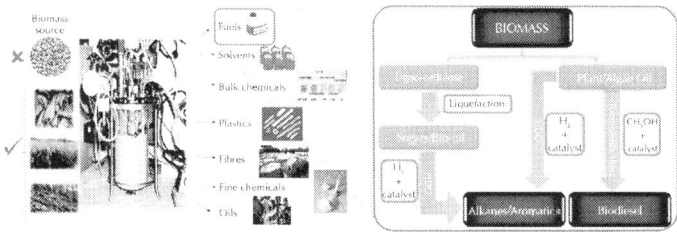

**Scheme 2:** Biorefinery routes for the co-production of chemicals and transportation fuels from biomass.

Biodiesel is a clean burning and biodegradable fuel which, when derived from non-food plant or algal oils or animal fats, is viewed as a viable alternative (or additive) to current petroleum-derived diesel.[22] Commercial biodiesel is currently synthesised *via* liquid base catalysed transesterification of $C_{14}$–$C_{20}$ triacylglyceride (TAG) components of lipids with $C_1$–$C_2$ alcohols[23-26] into fatty acid methyl esters (FAMEs) which constitute biodiesel as shown in Scheme 3, alongside glycerol as a potentially valuable by-product.[27] While the use of higher (e.g. $C_4$) alcohols is also possible,[28] and advantageous in respect of producing a less polar and corrosive FAME[29] with reduced cloud and pour points,[30] the current high cost of longer chain alcohols, and difficulties associated with separating the heavier FAME product from unreacted alcohol and glycerol, remain problematic. Unfortunately, homogeneous acid and base catalysts can corrode reactors and engine manifolds, and their removal from the resulting biofuel is particularly problematic and energy intensive, requiring aqueous quench and neutralisation steps which result in the formation of stable emulsions and soaps.[12,31,32] Such homogeneous approaches also yield the glycerine by-product, of significant potential value to the pharmaceutical and cosmetic industries, in a dilute aqueous phase contaminated by inorganic salts. The utility of solid base and acid catalysts for biodiesel production has been widely reported,[15,25,33-41] wherein they offer improved process efficiency by eliminating the need for quenching steps, allowing continuous operation,[42] and enhancing the purity of the glycerol by-product. Technical advances in catalyst and reactor design remain essential to utilise non-food based feedstocks, and thereby ensure that biodiesel remains a key player in the renewable energy sector for the 21st century. In this review, we highlight the contributions of tailored solid acid and base catalysts to catalytic biodiesel synthesis via TAG transesterification to FAMEs and free fatty acid (FFA) esterification.

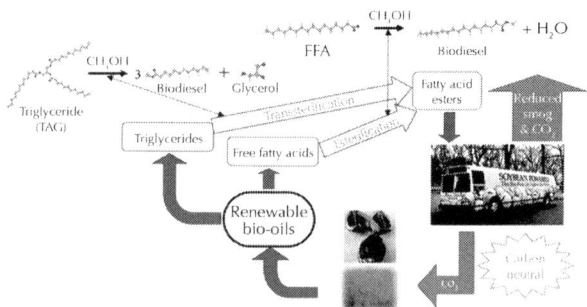

**Scheme 3:** Biodiesel production cycle from renewable bio-oils via catalytic transesterification and esterification.

# FEEDSTOCKS FOR BIODIESEL

The feedstock sources employed for biodiesel synthesis have remained little changed since the first engine tests with vegetable oils in the late 1800s,[43] and are normally classified as either first or second generation,[44,45] the latter oft referred to as a source of 'advanced biofuels'. First generation biodiesel is derived from edible vegetable oils such as soya, palm,[46] oil seed rape[47] and sunflower,[48] however the attendant poor yields (typically 3000–5000 L hectare$^{-1}$ year$^{-1}$) and socio-political concern over the diversion of such food crops for fuels has led to their fall from favour within Europe and North America. Second generation biodiesel is normally considered to be that obtained from non-edible oils such as castor,[49] Jatropha[50] and neem,[51] microalgae,[44,52] animal fats (e.g. tallow and yellow grease),[53] or waste oils including organic components of municipal waste:[54] these offer lower greenhouse gas emissions,[45] e.g. 150 $g_{CO2}$ MJ$^{-1}$ for African biodiesel from Jatropha exported to the EU with attendant use of residual seedcake as a fertiliser *versus* 220 $g_{CO2}$ MJ$^{-1}$ for Mexico biodiesel from Jatropha without attendant methane capture;[55] improved environmental and energy life cycles;[56] and superior biodiesel yields (upto 100 000 L hectare$^{-1}$ year$^{-1}$ for microalgae). Commercial biodiesel is require to meet a range of national and international standards, the most widely conformed to being the American standard ASTM D6751,[57] and the European standard EN 14214:[58] the high free fatty acid of some non-edible oils can lower the FAME content below accepted standards,[59] whereas feedstocks like *Brassica carinata* and *Jatropha curcas* have comparable or even higher oil content than many edible oils.[15]

Interest in biodiesel production soared following the global oil crisis of the 1970s, resulting in the United States, European Union, Brazil, China, India, and South Africa convening a UN International Biodiesel Forum for biodiesel development. Today, the United States, European Union and Brazil, alongside Malaysia, remain leading forces in the biodiesel market. Current industrial production is dominated by the utilisation of edible vegetable oils such as soybean (7.08 million), palm (6.34 million), rapeseed (6.01 million), castor, coconut and *Jatropha curcas* oil. The primary cost of biodiesel lies in the raw material, and since the market is dominated by food grade oils,[59] which are significantly more expensive than petroleum-derived diesel, economic viability remains to be proven. Use of the surplus from edible oil production may assist countries to meet the demands for biodiesel production without negatively impacting upon food requirements.[60] Feedstock selection is a strong function of local availability. Soybean oil, which is widely used in the United States and South America, is the third largest feedstock for biodiesel after rapeseed oil in Europe and palm oil in Asian countries, such as Malaysia and Indonesia, which also use sunflower and coconut oil, with *Jatropha curcas* oil widespread across South East Asia.[61] Soybean and rapeseed oils account for about 85% of global biodiesel production,[62] with 75% of total biodiesel produced in Europe. Competition for land to produce biodiesel feedstocks is problematic, hence maximising the yield of oil from a given feedstock is critical. Edible soybean seed consists of 20% oil *versus* rapeseed at 40%, whereas non-edible Jatropha and Karanja seeds contain around 40% and 33% oil respectively.[60] Adoption of soybean (as in the US) as a global biodiesel feedstock would be problematic, not only due to competition for its use as a food crop, but also the high quantities of waste, associated with its low oil yield, although this could be mitigated by the introduction of the oil seed cake as a major animal feed. The oil yield from non-edible Jatropha is particularly noteworthy since it can grow in poor quality soil and waste land, avoiding competition with arable land for food crops, however harvesting of the toxic seeds is labour intensive.[63] Around 15 million tons of waste cooking/frying oils is disposed of annually worldwide. Such low cost feedstocks, could meet a significant portion of current biodiesel demands, however chemical changes occurring during cooking which increase their FFA and moisture content must be taken into consideration.[64] Recent studies suggest that the production cost of biodiesel could be halved through waste cooking oils in comparison with virgin oils.[65] However because of its high melting point and viscosity, and less predictable supply, waste cooking oil has been less extensively investigated than vegetable oils.[31] Algal biomass has received considerable recent attention, since lipids from algae can be used for biodiesel production *via* conventional transesterification technologies. Microalgae are fast-growing and produce higher oil yields than plant counterparts. The high oil content of different microalgae favours their commercialisation as a promising feedstock: one acre of microalgae can produce 5000 gallons of biodiesel annually compared to only 70 gallons from an equivalent area of soybean,[52] and algae can flourish on land unusable for plant cultivation and without fresh water. Algal oil yields vary with the species, nutrient supply and harvest time,[66] however the properties of the resulting FAMEs are not superior to those derived from plant oils, and further research into algal oils rich in saturated long chain fatty acids is required in order to improve the quality of the final biodiesel.[67]

The choice of oil feedstock in turn influences the biodiesel composition and hence fuel properties,[43,68] notably acid value, oxidation stability, cloud point, cetane number and cold filter plugging point. Oils from plants usually comprise five major fatty acids components: palmitic (16 : 0); stearic (18 : 0); oleic (18 : 1); linoleic (18 : 2); and linolenic (18 : 3). Table 1 illustrates their distribution and associated physicochemical properties for some common feedstocks. High FFA oils not only compromise base catalysed transesterification and hence biodiesel yields, but can corrode engines and ancillary machinery; the acceptable acid range is between 0.5–3%.[60] The cetane number (CN), a measure of diesel ignition quality, is higher for biodiesel (46–52) than that of conventional diesel (40–55), with the international standard specified in ASTM D6751 and EN 14214 at 47 and 51 respectively. Cetane number varies with the degree of oil unsaturation and chain length. Esters of palmitic and stearic acid possess CNs higher than 80, while that of oleate is 55–58, with CN generally decreasing with increasing unsaturation (e.g. CN = 40 for linoleic and 25 for linolenic acid), falling to 48-5 for soybean- and 52–55 for rapeseed-derived biodiesel.[69] Fatty acid chain composition also influences $NO_x$ emissions, with biodiesel containing esters of saturated fatty acids emitting less $NO_x$ than petroleum diesel, and emissions increasing with the degree of unsaturation but decreasing with fatty acid chain length. $NO_x$ emissions of hydrogenated FAMEs derived from soybean oil is lower than from conventional diesel.[70]

**Table 1:** Common feedstocks for biodiesel production, free fatty acid composition and physicochemical properties. Reprinted from ref. 59, Copyright (2010), with permission from Elsevier

| | Feedstock | Composition/ wt% fatty acid | Density/g cm$^3$ | Flash point/°C | Acid value mg KOH g$^{-1}$ | Heating value/MJ kg$^{-1}$ |
|---|---|---|---|---|---|---|
| Edible oils | Soybean | C16:0, C18:1, C18:2 | 0.91 | 254 | 0.2 | 39.6 |
| | Rapeseed | C16:0, C18:0, C18:1, C18:2 | 0.91 | 246 | 2.92 | 39.7 |
| | Sunflower | C16:0, C18:0, C18:1, C18:2 | 0.92 | 274 | — | 39.6 |
| | Palm | C16:0, C18:0, C18:1, C18:2 | 0.92 | 267 | 0.1 | — |
| | Peanut | C16:0, C18:0, C18:1, C18:2, C20:0,C22:0 | 0.90 | 271 | 3 | 39.8 |
| | Corn | C16:0, C18:0, C18:1, C18:2, C18:3 | 0.91 | 277 | — | 39.5 |
| | Camelina | C16:0, C18:0, C18:1, C18:2, C18:3, C20:0, C20:1, C20:3 | 0.91 | — | 0.76 | 42.2 |
| | Cotton | C16:0, C18:0, C18:1, C18:2, C18:3 | 0.91 | 234 | — | 39.5 |

| Non-edible oils | Jatropha curcas | C16:0, C16:1, C18:0, C18:1, C18:2 | 0.92 | 225 | 28 | 38.5 |
|---|---|---|---|---|---|---|
| | Pongamina pinnata | C16:0, C18:0, C18:1, C18:2, C18:3 | 0.91 | 205 | 5.06 | 34 |
| | Palanga | C16:0, C18:0, C18:1, C18:2 | 0.90 | 221 | 44 | 39.25 |
| | Tallow | C14:0, C16:0, C16:1, C17:0, C18:0, C18:1, C18:2 | 0.92 | — | — | 40.05 |
| | Poultry | C16:0, C16:1, C18:0, C18:1, C18:2, C18:3 | 0.90 | — | — | 39.4 |
| | Used cooking oil | Depends on fresh cooking oil | 0.90 | — | 2.5 | — |

Oxidation stability also depends upon the degree of unsaturation of fatty acid chains within the oil feedstock, since double bonds are prone to oxidation. Biodiesel produced from feedstocks containing linoleic (C18, two C=C double bonds) and linolenic acid (C18, three C=C double bonds), with one or two bis-allylic positions, are highly susceptible to oxidation. The relative rates of oxidation for linoleates and linolenates are respectively 41 and 98 times higher than that of the monounsaturated oleate.[71] The viscosity of biodiesel also increases with chain length and saturation of fatty acids within the feedstock,[72] influencing the fuel lubricity and flow properties. Low viscosity biodiesel can be obtained from low molecular weight triglycerides, however such biodiesel cannot be used directly as a fuel due to its poor cold temperature flow properties. The kinematic viscosities of the two most common biodiesels are 4.0–4.1 mm$^2$ s$^{-1}$ from soybean oil and 4.4 mm$^2$ s$^{-1}$ from rapeseed oil. The lubricity of biodiesel increases with chain length, and the presence of double bonds and alcohol groups. Hence, monoglycerides and trace glycerol increase biodiesel lubricity. The high lubricity of biodiesel can be utilised through blending with conventional, low-sulfur diesel to improve overall fuel lubricity.[73] Cold point (CP) and pour point (PP) determine the flow properties of biodiesel, and also depend on the fatty acid composition of the feedstock. CP is the temperature at which a fuel begins to solidify, and PP is the temperature at which the fuel can no longer flow. For conventional diesel, CP and PP values are −16 °C and −27 °C respectively. Biodiesel derived from soybean possesses CP and PP values of around 0 °C to −2 °C, while the CP for rapeseed oil-derived biodiesel is −3 °C. These values are very high in comparison to conventional diesel, rendering biodiesel ill-suited for cold countries.[70] Other common feedstocks, such as palm oil, jatropha oil, animal fat and waste cooking oil have even higher CP values of around 15 °C. In contrast, biodiesel derived from *cuphea* oil enriched with saturated, medium-chain C8–C14 fatty acids exhibits improved properties including a lower CP of −9 to −10 °C,[74] comparable to conventional diesel. Genetic engineering of the parent plants or microalgae offers a route to optimise the fatty acid composition of feedstock oils to deliver fuels with the desired physicochemical properties.[75]

# SOLID BASE CATALYSED BIODIESEL SYNTHESIS

Base catalysts are generally more active than acids in transesterification, and hence are particularly suitable for high purity oils with low FFA content. Biodiesel synthesis using a solid base catalyst in continuous flow, packed bed arrangement would facilitate both catalyst separation and co-production of high purity glycerol, thereby reducing production costs and enabling catalyst re-use. Diverse solid base catalysts are known, notably alkali or alkaline earth oxides, supported alkali metals, basic zeolites and clays such as hydrotalcites, and immobilised organic bases.[76]

## Alkaline earth oxides

Basicity in alkaline earth oxides is believed to arise from $M^{2+}$–$O^{2-}$ ion pairs present in different coordination environments.[77] The strongest base sites occur at low coordination defect, corner and edge sites, or on high Miller index surfaces. Such classic heterogeneous base catalysts have been extensively tested for TAG transesterification[78] and there are numerous reports on commercial and microcrystalline CaO applied to rapeseed, sunflower or vegetable oil transesterification with methanol.[79,80] Promising results have been obtained, with 97% oil conversion achieved at 75 °C,[80] however concern remains over

$Ca^{2+}$ leaching under reaction conditions and associated homogeneous catalytic contributions,[81] a common problem encountered in metal catalysed biodiesel production which hampers commercialisation.[82] While Ca and Mg are the more widely used alkaline earth metals in solid base catalysis, strontium oxides have also found application in biodiesel production. Pure strontium oxide possesses the highest base site density of the alkali earth oxides as determined by $CO_2$ temperature programmed desorption (TPD),[83] and a comparable base strength to that of BaO ($26.5 < H_-$). Despite the lower surface area of SrO compared to Mg and Ca oxides (19, 14 and 3 $m^2$ $g^{-1}$ respectively), it showed the highest activity for hempseed oil transesterification, although it is questionable whether such low area/highly soluble materials could ever be commercially viable.

Alkali-doped CaO and MgO have also been investigated for TAG transesterification,[84–86] with their enhanced basicity attributed to the genesis of $O^-$ centres following the replacement of $M^+$ for $M^{2+}$ and associated charge imbalance and concomitant defect generation. In the case of Li-doped CaO, the electronic structure of surface lithium ions (as probed by XPS) evolves discontinuously as a function of concentration and phase. Maximal activity was observed upon formation of a saturated $Li^+$ monolayer, with the phase to bulk-like $LiNO_3$ at higher loadings suppressing TAG conversion coincident with loss of strong base sites.[86] However, leaching of alkali promoters remains problematic.[87]

It is widely accepted that the catalytic activity of alkaline earth oxide catalysts is very sensitive to their preparation, and corresponding surface morphology and/or defect density. For example, Parvulescu and Richards demonstrated the impact of the different MgO crystal facets upon the transesterification of sunflower oil by comparing nanoparticles[88] versus (111) terminated nanosheets.[89] Chemical titration revealed that both morphologies possess two types of base sites, with the nanosheets exhibiting well-defined, medium-strong basicity consistent with their uniform exposed facets and which confer higher FAME yields during sunflower oil transesterification (albeit scale-up of the nanosheet catalyst synthesis may be costly and non-trivial). Subsequent synthesis, screening and spectroscopic characterisation of a family of size-/shape-controlled MgO nanoparticles prepared via a hydrothermal synthesis, revealed small (<8 nm) particles terminate in high coordination (100) facets, and exhibit both weak polarisability and poor activity in tributyrin transesterification with methanol.[90] Calcination drives restructuring and sintering to expose lower coordination stepped (111) and (110) surface planes, which are more polarisable and exhibit much higher transesterification activities under mild conditions. A direct correlation was therefore observed between the surface electronic structure and associated catalytic activity, revealing a pronounced structural preference for (110) and (111) facets (Fig. 1). *In situ* aberration corrected-transmission electron microscopy and XPS implicates coplanar anion vacancies as the active sites in tributyrin transesterification with the density of surface defects predicting activity.[90,91]

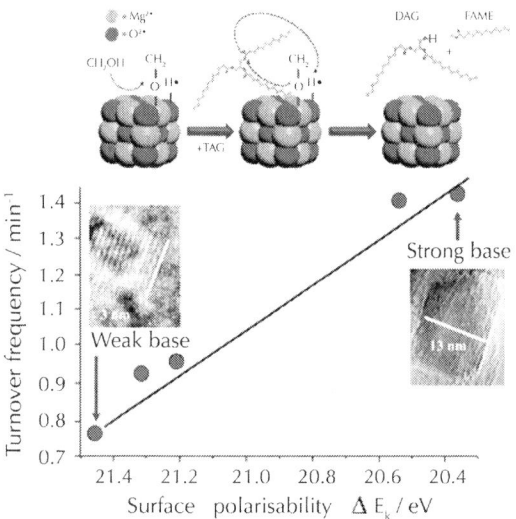

**Figure 1:** Relationship between surface polarisability of MgO nanocrystals and their turnover frequency towards tributyrin transesterifcation. Adapted from ref. 90 with permission from The Royal Society of Chemistry.

Cesium doping *via* co-precipitation under supercritical conditions confers even greater activity towards tributyrin transesterification with methanol,[85] due to the genesis of additional, and stronger, base sites associated with a new ordered mixed oxide phase which EXAFS analysis recently identified as $Cs_2Mg(CO_3)_2(H_2O)_4$,[92] resulting in superior performance compared with MgO and even homogeneous $Cs_2CO_3$ catalysts (Fig. 2). Unfortunately, surface carbon deposition and loss of this high activity $Cs_2Mg(CO_3)_2(H_2O)_4$ phase due to partial

Cs dissolution results in on-stream deactivation of Cs-doped MgO, although recalcination could help to regenerate activity.

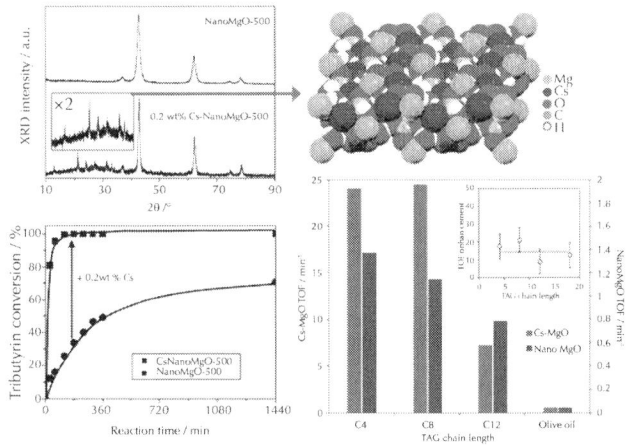

**Figure 2:** Formation of crystalline $Cs_2Mg(CO_3)_2(H_2O)_4$ phase within co-precipitated Cs-doped MgO and resulting synergy in the transesterification of short and long chain TAGs with methanol compared with undoped nanocrystalline MgO. Adapted from ref. 85 with kind permission from Springer Science and Business Media and ref. 92 with permission from John Wiley and Sons.

Alkaline earth metal oxides may be incorporated into metal oxides to form composite oxides[93] which are also suitable as solid base catalysts for biodiesel production. The activity of such composites is similar to that of the parent alkaline earth (typically CaO), but they exhibit greater stability and are less prone to dissolution, facilitating separation from the reaction media. Calcination temperature strongly influences the resulting catalytic activity towards transesterification. For example, a Ca–Al composite oxide containing $Ca_{12}Al_{14}O_{33}$ and CaO thermally processed between 120 °C and 1000 °C showed maximal activity after a 600 °C treatment due to changes in specific surface area and crystallinity. CaO was only observed in samples prepared >600 °C, accompanied by the formation of crystalline $Ca_{12}Al_{14}O_{33}$. Synergy between these two phases greatly improved the transesterification activity, however calcination at temperatures significantly above 600 °C induced crystallite sintering and concomitant loss of surface area and activity. Unfortunately the catalyst synthesis employed sodium precursors, hence alkali contamination of these catalysts cannot be discounted, and which in any event were employed at high loadings (6 wt%) and without recycle tests.

Calcium also forms a mixed oxide with $MoO_3$.[94] Supporting both oxides on SBA-15 mesoporous silica afforded a transesterification catalyst with improved stability relative to CaO due the presence of acidic $MoO_3$ sites on the SBA-15. The impact of Ca : Mo ratio and calcination temperatures was explored, with a Ca : Mo ratio of 6 : 1 maximising activity for soybean oil conversion, boosting FAME yields from 48 to 83% over extremely long reaction times in excess of 50 h. Raising the calcination temperature from 350 °C to 550 °C induced CaO and $MoO_3$ crystallisation, with a corresponding rise in activity; higher temperature calcination did not promote further crystallisation and was not beneficial for transesterification.

Alkaline earth oxides may be used to support acidic or amphoteric materials to form materials with mixed acid–base character. Transesterification of soybean oil over CaO supported $SnO_2$ prepared via impregnation was highly dependent on calcination temperature and the Ca : Sn ratio.[95] The interaction between acidic $SnO_2$ and basic CaO resulted in a highly $SnO_2$ phase and associated active sites. Calcination above 350 °C was required to initiate decomposition of the Ca precursor, with temperatures >650 °C driving complete conversion to Ca oxides. Optimal performance was obtained for high calcination temperatures, which maximised the CaO content. Further heating again led to particle sintering/agglomeration and decreased reactivity. Supported CuO can also produce biodiesel from hempseed oil,[83] with 10 wt% CuO/SrO offering 20% higher FAME yields under optimised conditions than other alkaline earth oxides. The CuO could also undergo chemical reduction during transesterification to form an active catalyst for the selective hydrogenation of polyunsaturated hydrocarbons for further biodiesel upgrading. It should be noted that the catalyst loadings employed in this study of 4–12 wt% would likely prove prohibitive in any commercial process, and that small but significant (29 ppm) quantities of leached Ca may have contributed to the observed performance.

Composites of Sr and Al were prepared by Farzaneh et al. and evaluated for soybean oil transesterification with methanol.[96] The dominant crystalline phase was $Sr_3Al_2O_6$, giving rise to medium and high strength base sites with corresponding $CO_2$ desorption peak maxima of 388 °C and 747 °C respectively. The Sr–Al oxide also possessed a higher density of base sites compared to solid bases such as $CaO/Al_2O_3$, reflected in an eight-fold higher $CO_2$ adsorption capacity. These superior base properties enhanced the activity of the strontium composite for soybean transesterification to FAMEs, resulting in comparable conversions at a lower catalyst loading and shorter reaction time than for a MgAl hydrotalcite and $CaO/Al_2O_3$. While oil conversions fell noticeably with repeated re-use, there was no evidence of alkaline earth dissolution, and the resulting biodiesel fuel met ASTM and EN standards.

## Alkali Doped Materials

As shown in Fig. 1, lithium doped CaO can enhance tributyrin transesterification. Li doping has also been exploited over $SiO_2$, wherein 800 °C calcination results in a lithium orthosilicate solid base catalyst, $Li_4SiO_4$.[97] Although the basic strength of $Li_4SiO_4$, determined by Hammett indicators, was less than that of CaO, both materials exhibited similar initial activity towards soybean transesterification, with the lithium orthosilicate more stable and maintaining activity after prolonged exposure to air, in contrast to CaO. The superior stability of the $Li_4SiO_4$ catalyst was further demonstrated by its water and carbon dioxide tolerance, both of which poison conventional alkaline earth catalysts.

Sodium silicate, $Na_2SiO_3$, is also active for biodiesel production from rapeseed and *jatropha* oils under both conventional[98] and microwave assisted conditions,[99] with a 98% FAME yield after one hour reaction under mild conditions. Although this catalyst displayed good recyclability, TAG conversions fell steadily to <60% after four re-uses, attributed to water adsorption and Si–O–Si bond cleavage and sodium leaching.[98] The same catalyst was evaluated using microwave heating for only five minutes at a range of powers between 100–500 W (Fig. 3).[99] At low power only 18% rapeseed oil conversion was obtained. Higher powers heated the reaction mixture (to ~175 C for 400 W) in turn boosting FAME yields from both oils to ~90%, highlighting the use of microwave heating to accelerate biodiesel production. Recycle studies again showed slow *in situ* deactivation due to particle agglomeration, water adsorption of water, and associated loss of basicity due to sodium leaching into methanol during both transesterification and washing procedures between recycles. Despite some recent successes in the scale-up of microwave-assisted (homogeneously catalysed) biodiesel production (see Section 6),[28,100] it remains unlikely that such heating solutions can deliver the high throughput demanded for commercial processes.

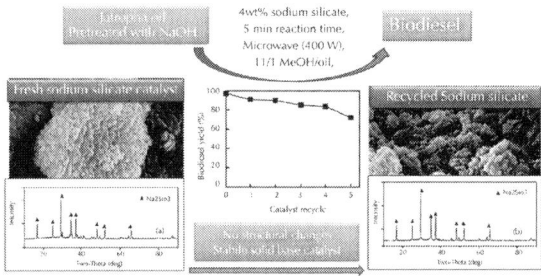

**Figure 3:** Demonstration of the structural stability and catalytic activity of sodium silicate as a solid base for biodiesel production. Adapted from ref. 99. Copyright (2014), with permission from Elsevier.

Activated carbon can be used as an amphoteric support for basic alkaline metal salts such as $K_2CO_3$,[101] which is known to be an active homogeneous catalyst for oil transesterification and biodiesel production.[102] A study of $K_2CO_3$ supported over a range of support materials, such as MgO, activated carbon and $SiO_2$, demonstrated that $K_2CO_3$ on basic carriers gave higher activity for rapeseed oil transesterification than when using acidic carriers (unsurprisingly due to self-neutralisation!).[102] $K_2CO_3$/MgO was shown to be highly stable, with spent catalysts showing minimal loss of performance over six re-uses (though requiring 400 °C reactivation between cycles), and exhibiting negligible structural changes or potassium leaching. Kraft lignin is a low cost, renewable by-product of the Kraft wood pulping process, and possesses high carbon and low ash content and is therefore a popular precursor for activated carbons. Li *et al.* used $K_2CO_3$ in a one-pot method to prepare activated carbon and transform this into a solid base catalyst, namely $K_2CO_3$ on Kraft Lignin activated carbon (LKC), for biodiesel production.[101] Thermal activation had a significant impact on the resulting catalytic activity, with higher calcination temperatures increasing the surface area and pore volume 100-fold and hence FAME production, however temperatures above 800 °C induced $K_2CO_3$

decomposition and poorer performance. Optimal reaction conditions of 65 °C, 3 wt% loading and a K/KLC ratio of 0.6, enabled a 98% FAME yield from rapeseed oil transesterification, which fell to 82% after four recycles as a result of progressive particle agglomeration and potassium leaching into the biodiesel. Wu *et al.* supported a range of potassium salts on mesoporous silicas for use as solid base biodiesel catalysts.[103] A $K_2SiO_3$ impregnated catalyst proved superior to $K_2CO_3$ and KAc impregnated catalysts due to its higher base site density (1.94 *versus* 1.81 and 1.72 mmol g$^{-1}$ respectively). Aluminium addition to the SBA-15 framework improved the morphology, increasing the surface area and pore volume, and $CO_2$ desorption temperature indicative of a more strongly basic support; this observation is rather counter-intuitive, since Al-doping of SBA-15 is usually employed to promote the formation of Brönsted and Lewis acid sites of moderate acidity.[104] A 30% $K_2SiO_3$/ AlSBA-15 catalyst was used for the transesterification of Jatropha oil with MeOH at 60 °C, giving 95% conversion for a relatively low MeOH/ oil molar ratio of 9 : 1. This catalyst was recycled five times with only a 6% drop in conversion, but the filtered catalyst required regenerative washing with a methanol–*n*-hexane mixture and re-calcination to avoid a significant drop in FAME yield to 47% after the fifth recycle. The magnitude of this activity loss indicates significant K leaching. In a related study, Xie *et al.* immobilised tetraalkylammonium hydroxides onto SBA-15 for soybean oil transesterification.[105] The resulting SBA-15-pr-NR$_3$OH catalyst gave 99% conversion to FAMEs under methanol reflux. Covalent linking of the tetraalkylammonium hydroxide to the silica surface prevented *in situ* leaching, resulting in only a 1% fall in FAME yield after five recycles and appears a promising methodology for biodiesel production at mild-moderate temperatures under which the covalently linked propyl backbone is thermally stable.

Despite its importance in the context of second generation biofuels, waste biomass has been less extensively investigated in catalyst preparation. Most such studies have focused on the synthesis of carbonaceous solid acid catalysts[2,106–109] as discussed later. In contrast, rice husk ash modified with Li *via* a simple solid state preparative route, has been exploited as a solid base catalyst by for soybean oil transesterification with methanol.[106] These materials exhibited high basicity (H_ > 15.0), comparable to that of CaO, and consequent high activity, but superior air stability than CaO which deactivated due to hydration; the Li rice husk catalyst showed only a modest drop in oil conversion from 97% to 82% upon re-use. As with any material derived from a biogenic source the question of compositional variability arises, particularly in regard to residual heavy metals in the ash, which is likely to hamper catalyst reproducibility.[110]

# Transition Metal Oxides

Solid bases usually afford higher rates of transesterification than solid acids, hence a range of transition metal oxides of varying Lewis base character have been explored in biodiesel production. MnO and TiO are mild bases with good activity for biodiesel production,[111] and have been applied for the simultaneous transesterification of triglycerides and esterification of FFAs under continuous flow conditions using low grade feedstocks with high fatty acid contents (up to 15%). Soap formation, caused by leaching of metal from the catalyst surface under high FFA concentrations, was an order of magnitude less than that observed with conventional homogeneous base catalysts. Unfortunately, this study did not characterise the Mn or Ti oxidation state in either fresh or spent materials to confirm the nature of any catalytic centre. Zirconium has also been shown to activate and stabilise solid base catalysts for biodiesel production.[101,112,113] Mixed oxides of CaO and $ZrO_2$ prepared via co-precipitation showed increased surface area and stability with increasing Zr : Ca ratios (Fig. 4). However, the transesterification activity remained dependent upon the Ca content, decreasing at lower CaO loadings.[112] Sodium zirconate, a potential $CO_2$ adsorbent,[84,114] has shown promise in biodiesel production,[113] with 98% conversion of soybean oil to FAME after 3 h at 65 °C. Deactivation observed upon repeated decanting and recycling was attributed to surface poisoning, with methanol washing between cycles facilitating 84% conversion after five recycles. This material›s affinity for carbon dioxide and large crystallite size/low surface area (~1 $m^2$ $g^{-1}$) may render it air-sensitive and prone to further sintering. Zirconia was employed as a support for a range of sodium-containing bases, such as NaOH, $NaH_2PO_4$, $C_4H_5O_6Na$ (monosodium tartrate) and potassium sodium tartrate were doped on $ZrO_2$ to prepare a series of catalysts with varying basic strength and total basicity for the microwave assisted transesterification of soybean oil with methanol.[101] Catalytic activity was dependent upon basicity, increasing at higher Na : Zr ratios. The potassium sodium tartrate doped zirconia exhibited the strongest basicity and highest conversions, reaching 54% for Na : Zr = 1 and a 1 : 10 catalyst : soybean oil mass ratio at 60 °C under 600 W microwave power. Increasing the Na : Zr ratio to 2 improved conversion to 92%. Optimal conversions were obtained for catalysts calcined at 600 °C, possibly due to tartrate decomposition at higher temperatures, although this catalyst was recyclable*via* filtration and re-calcination.

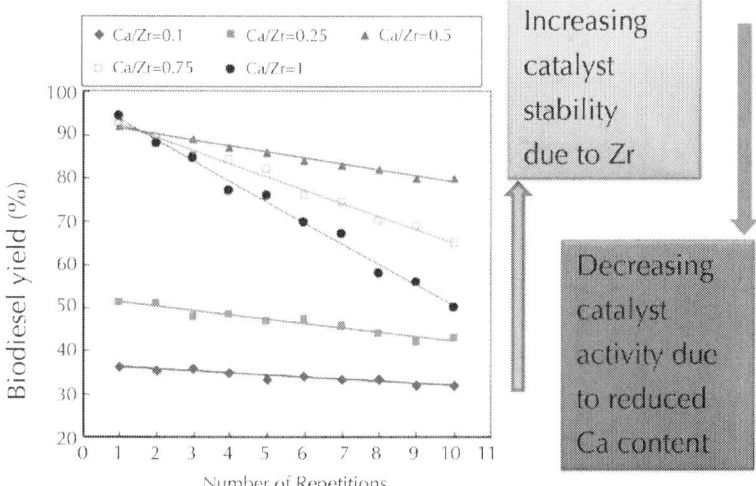

**Figure 4:** Effect of Zr-doping on CaO solid base catalysts for biodiesel production. Adapted from reference 112. Copyright (2012), with permission from Elsevier.

Porosity was introduced to a titania-based catalyst through the construction of sodium titanate nanotubes as solid base catalysts for soybean oil transesterification with methanol.[115] The catalyst exhibited a range of active sites of varying basicity, however the high sodium content (10 wt%) is a cause for concern due to the high probability of leaching *in situ* and associated homogeneous chemistry. The pore distribution was bimodal, consisting of 3 nm wide tubular mesopores and ~40 nm voids between the aggregated nanotubes. Biodiesel yields of >97% were obtained for 1–2 wt% of catalyst at 65 °C. However, a large excess of methanol to oil was required (40 : 1 molar ratio), and while this material could be re-used several times, it was less active than that of CaO and MgO lacking such a nanoporous architecture.

# Hydrotalcites

Hydrotalcites are another class of solid base catalysts that have attracted attention because of their high activity and robustness in the
presence of water.[116,117] Hydrotalcites ($[M(II)_{1-x}M(III)_x(OH)_2]^{x+}(A^{n-}_{x/n})\cdot mH_2O$) adopt a layered double hydroxide structure with brucite-like
($Mg(OH)_2$) hydroxide sheets containing octahedrally coordinated $M^{2+}$
and $M^{3+}$ cations, separated by interlayer $A^{n-}$ anions to balance the overall charge,[118] and are conventionally synthesised *via* co-precipitation
from their nitrates using alkalis as both pH regulators and a carbonate
source. Mg–Al hydrotalcites have been applied to TAG transesterification of poor and high quality oil feeds,[119] such as refined and acidic
cottonseed oil (possessing 9.5 wt% FFA) and animal fat feed (45 wt%
water), delivering 99% conversion within 3 h at 200 °C. It is important
to note that many catalytic studies employing hydrotalcites for transesterification are suspect due to their use of Na or K hydroxide/carbon-
ate solutions to precipitate the hydrotalcite phase. Complete removal
of alkali residues from the resulting hydrotalcites is inherently difficult,
resulting in ill-defined homogeneous contributions to catalysis arising
from leached Na or K.[120,121] This problem has been overcome by the
development of alkali-free precipitation routes employing $NH_3OH$ and
$NH_3CO_3$, which offer well-defined, thermally activated and rehydrat-
ed Mg–Al hydrotalcites with compositions spanning $x = 0.25$–$0.55$.[116]
Spectroscopic measurements reveal that increasing the Mg : Al ratio
enables systematic enhancement of the surface charge and accompa-
nying base strength, with a concomitant increase in the rate of tributyrin transesterification under mild reaction conditions (Fig. 5). Despite
their high intrinsic activity, one limitation of co-precipitated pure hydrotalcites is their low surface areas, although delamination[122,123] and
grafting[124] methodologies offer avenues to circumvent this.

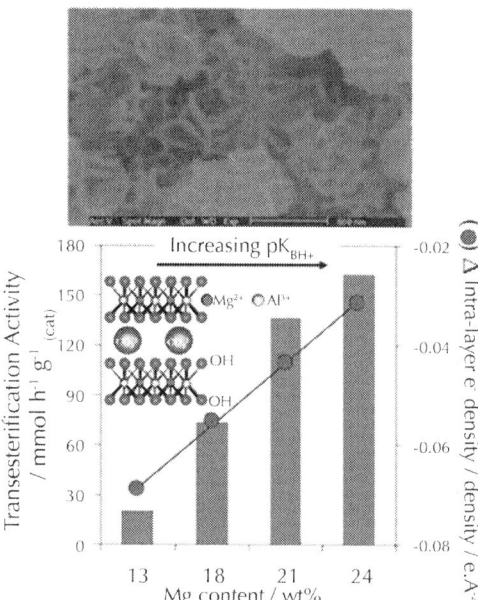

**Figure 5:** Impact of Mg:Al hydrotalcite surface basicity on their activity towards tributyrin transesterification. Adapted from ref. 117. Copyright (2005), with permission from Elsevier.

Since conventionally-prepared hydrotalcites are microporous, they are poorly suited to transesterification of bulky $C_{16}$–$C_{18}$ TAGs which are the principal components of bio-oils. One solution has therefore been to utilise catalysts possessing a bimodal pore distribution, wherein micropores provide a high surface density of base sites while a complementary meso- or macropore network affords rapid transport of TAGs from the bulk reaction media to these active sites, and removal of FAME and glycerol products back out from the porous catalyst. Ordered, hierarchical materials possessing such bimodal pore architectures can be prepared by combining hard and soft templating approaches, exemplified by the methodology developed by Géraud and co-workers, wherein co-precipitation of the divalent and trivalent metal cations occurs within the interstices of an infiltrated polystyrene (PS) colloidal crystal.[125,126] This approach has been adopted to incorporate macroporosity into an alkali-free Mg–Al hydrotalcite, and thus create a hierarchical macroporous–microporous hydrotalcite solid base catalyst.[127] The resulting macropores act as rapid access conduits to transport heavy TAG oil components to active base sites present at the surface of (high aspect ratio) hydrotalcite nanocrystallites, thereby promoting triolein transesterification compared with that achievable over a Mg–Al microporous hydrotalcite of identical chemical composition (Fig. 6). Spiking experiments confirm that transesterification of the bulky $C_{18}$ triolein by the hierarchical hydrotalcite catalyst is less hindered by reactively-formed glycerol than when using a conventional microporous hydrotalcite (wherein glycerol completely suppresses biodiesel production). In contrast the more mobile model $C_4$ TAG, trubutyrin possesses an infinite dilution diffusion coefficient of 0.074 $cm^2$ $s^{-1}$ in methanol *versus* 0.037 $cm^2$ $s^{-1}$ for the triolein in methanol. Future scalability of such hierarchical catalysts will require either improved extraction protocols to enable re-use of the colloidal PS template, or the development of alternative polymeric templates derived from sustainable resources, such as polylactic or poly(lactide-*co*-glycolide) nanospheres.[128]

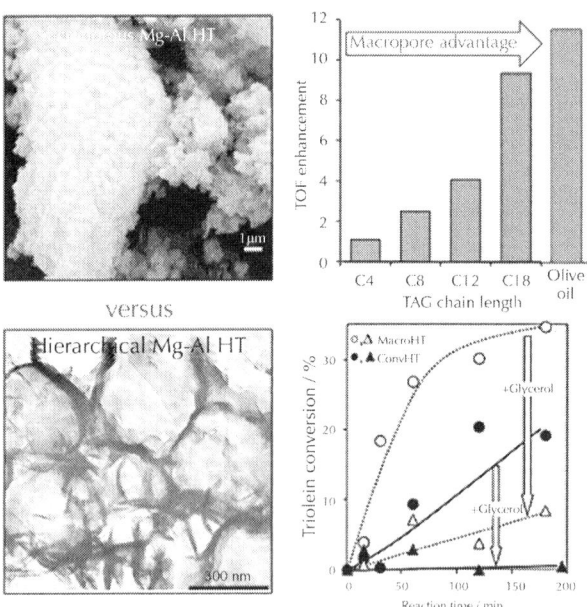

**Figure 6:** Superior catalytic performance of a hierarchical macroporous–microporous Mg–Al hydrotalcite solid base catalyst for TAG transesterification to biodiesel *versus* a conventional microporous analogue. Adapted from ref. 128 with permission from The Royal Society of Chemistry.

In terms of sustainability, it is important to find low cost routes to the synthesis of solid base catalysts that employ earth abundant elements. Dolomitic rock, comprising alternating $Mg(CO_3)$–$Ca(CO_3)$ layers, is structurally very similar to calcite ($CaCO_3$), with a high natural abundance and low toxicity, and in the UK is sourced from quarries working Permian dolomites in Durham, South Yorkshire and Derbyshire.[129] In addition to uses in agriculture and construction, dolomite finds industrial applications in iron and steel production, glass manufacturing and as fillers in plastics, paints, rubbers, adhesives and sealants. Catalytic applications for powdered, dolomitic rock offer the potential to further valorise this readily available waste mineral, and indeed dolomite has shown promise in biomass gasification[130] as a cheap, disposable and naturally occurring material that significantly reduces the tar content of gaseous products from gasifiers. Dolomite has also been investigated as a solid base catalyst in biodiesel synthesis,[131] wherein fresh dolomitic rock comprised approximately 77% dolomite and 23% magnesian calcite. High temperature calcination induced Mg surface segregation, resulting in MgO nanocrystals dispersed over $CaO/(OH)_2$ particles, while the attendant loss of $CO_2$ increases both the surface area and basicity. The resulting calcined dolomite proved an effective catalyst for the transesterification of $C_4$, $C_8$ and TAGs with methanol and longer chain $C_{16-18}$ components present within olive oil, with TOFs for tributyrin conversion to methyl butanoate the highest reported for any solid base. The slower transesterification rates for bulkier TAGs were attributed to diffusion limitations in their access to base sites. Calcined dolomite has also shown promise in the transesterification of canola oil with methanol, achieving 92% FAME after 3 h reaction with 3 wt% catalyst.[132]

Doping of (calcined) Malaysian dolomite with ZnO and $SnO_2$ resulted in respective three- and four-fold increases in the catalyst surface area and active base density, and a concomitant rise in base strength.[133] The $SnO_2$ doped dolomite gave >99.9% conversion under optimised conditions with a low methanol : oil molar ratio and catalyst loading.

Other waste materials employed for biodiesel production include waste water scale (obtained from residential kitchens in China), which upon 1000 °C calcination yielded a solid base material mixture of CaO, MgO, $Fe_2O_3$, $Al_2O_3$, and $SiO_2$ as a stable and active catalyst for soybean transesterification with methanol.[134] This composition is similar to that of Red Mud mineral waste, recently shown to be an active ketonisation catalyst.[135,136] This waste to resource approach of catalyst design is highly desirable in terms of green credentials and the biofuel ideology.

In summary, a host of inorganic solid base catalysts have been developed for the low temperature transesterification of triglyceride components of bio-oil feedstocks, offering activities far superior to those achieved *via* alternative solid acid catalysts to date. However, leaching of alkali and alkaline-earth elements and associated catalyst recycling remains a challenge, while improved resilience to water and fatty acid impurities in plant, algal and waste oil feedstocks is required in order to eliminate additional esterification pre-treatments.

# SOLID ACID CATALYSED BIODIESEL SYNTHESIS

A wide range of inorganic and polymeric solid acids are commercially available, however their application for the transesterification of oils into biodiesel has only been recently explored, in part reflecting their lower activity compared with base-catalysed routes,[32] in turn necessitating higher reaction temperatures to deliver suitable conversions. Despite their generally poorer activity, solid acids have the advantage that they are less sensitive to FFA contaminants then their solid base analogues, and hence can operate with unrefined feedstocks containing high acid contents.[32] In contrast to solid bases, which require feedstock pretreatment to remove these fatty acid impurities, solid acids are able to esterify FFAs through to FAME in parallel with transesterification of major TAG components, without saponification, and hence enable a reduction in the number of processing steps to biodiesel.[137–139]

## Mesoporous silicas

Mesoporous silicas from the SBA family[140] have been examined for biodiesel synthesis, and include materials grafted with sulfonic acid groups[141,142] or $SO_4/ZrO_2$ surface coatings.[143] Phenyl and propyl sulfonic acid SBA-15 catalysts are particularly attractive materials with activities comparable to Nafion and Amberlyst resins in palmitic acid esterification.[144] Phenylsulfonic acid functionalised silica are reportedly more active than their corresponding propyl analogues, in line with their respective acid strengths, but are more difficult to prepare. Unfortunately, conventionally synthesised sulfonic acid-functionalised SBA-15 silicas possess pore sizes below ~6 nm and long, isolated parallel channels, and suffer correspondingly slow in-pore diffusion and catalytic turnover in FFA esterification. However, poragens such as trimethylbenzene,[145] triethylbenzene and triisopropylbenzene[146] can induce swelling of the Pluronic P123 micelles used to produce SBA-15, enabling ordered mesoporous silicas with diameters spanning 5–30 nm. This methodology was recently applied to prepare a range of large pore SBA-15 materials employing trimethylbenzene as the poragen, resulting in the formation of highly-ordered periodic mesostructures with pore diameters of ~6, 8 and 14 nm.[127] These silicas were subsequently functionalised by mercaptopropyl trimethoxysilane (MPTS) and oxidised with $H_2O_2$ to yield expanded $PrSO_3$-SBA-15 catalysts which were effective in both palmitic acid esterification with methanol and tricaprylin and triolein transesterification with methanol under mild conditions. For both reactions, turnover frequencies dramatically increased with pore diameter, and all sulfonic acid heterogeneous catalysts significantly outperformed a commercial Amberlyst resin. These rate enhancements are attributed to superior mass-transport of the bulky free fatty acid and triglycerides within the expanded $PrSO_3$-SBA-15. Similar observations have been made over poly(styrenesulfonic acid)-functionalised, ultra-large pore SBA-15 in the esterification of oleic acid with butanol.[147] Mesopore expansion accelerates reactant/product diffusion to/from active sites, but there are limits to the extent to which this can be achieved without concomitant loss of pore ordering, which hampers mesoscopic modelling.[148]

The two dimensional, micron-length channels characteristic of the SBA-15 *p6mm* structure are known to hamper rapid molecular exchange with the bulk reaction media, and hence three dimensional interconnected channels associated with the *Ia3d* structure of KIT-6 mesoporous silica offer one solution to improving the in-pore accessibility of sulfonic acid sites. Superior molecular transport within the interconnected cubic structure of KIT-6 has been shown to facilitate biomolecule immobilisation.[149] This diversity of mesoporous silica architectures enabled the impact of pore connectivity upon FFA esterificationtobequantified.[150]Afamilyofpore-expandedpropylsulfonic acid KIT-6 analogues, PrSO$_3$H-KIT-6, prepared *via* MPTS grafting and subsequent oxidation, have been screened for FFA esterification with methanol under mild conditions. Such a conventionally-prepared material exhibited 40 and 70% TOF enhancements for propanoic and hexanoic acid esterification respectively over an analogous PrSO$_3$H-SBA-15 catalyst of comparable (5 nm) pore diameter, attributed to faster mesopore diffusion. However, pore accessibility remained rate-limiting for esterification of the longer chain lauric and palmitic acids. Pore expansion of the KIT-6 mesopores up to 7 nm *via* hydrothermal ageing doubled the resulting TOFs for lauric and palmitic acid esterification with respect to an unexpanded PrSO$_3$H-SBA-15 (Fig. 7). It should be noted that the absolute conversions of FFAs over such tailored, inorganic solid acid catalysts remain significantly lower than those for commercial polymer alternatives which possess superior acid site densities (*e.g.* 4.7 mmol g$^{-1}$ for Amberlyst-15[151] *versus* <1 mmol g$^{-1}$ for PrSO$_3$H-SBA-15 and PrSO$_3$H-KIT-6[150]).

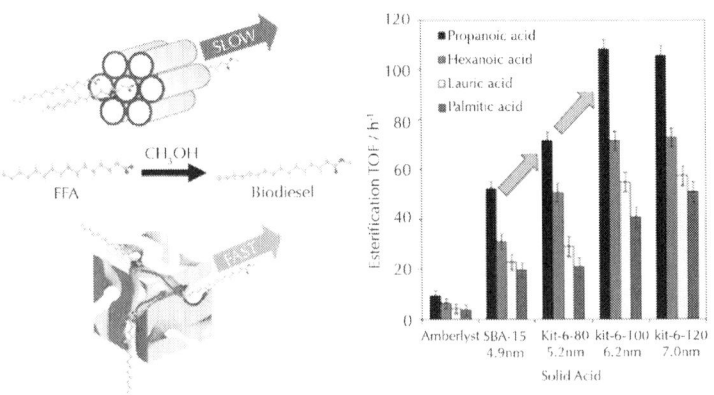

**Figure 7:** Superior performance of interconnected, mesoporous propylsulfonic acid KIT-6 catalysts for biodiesel synthesis *via* free fatty acid esterification with methanol *versus* non-interconnected mesoporous SBA-15 analogue. Adapted from ref. 151. Copyright 2012 American Chemical Society.

Propylsulfonic acid functionalised SBA-15 (SBA-15-PrSO$_3$H) has also been evaluated for oleic acid esterification with methanol,[152] showing good stability in boiling water, with the mesopore structure allowing facile diffusion of the acid to active sites. This catalyst exhibited similar activity to phenylethylsulfonic acid functionalised silica gel, and was superior to dry Amberlyst-15, reflecting the higher surface area and pore volume of the SBA-15-PrSO$_3$H relative to the more strongly acidic phenylethyl mesoporous silica. The SBA-15-PrSO$_3$H could be recycled by simple ethanol washing and drying at 80 °C, and maintained an esterification rate of 2.2 mmol min$^{-1}$ g$_{cat}$$^{-1}$. Simultaneous esterification and transesterification of vegetable oils with methanol has performed with Ti-doped SBA-15.[153] A range of oils including soybean, rapeseed, crude palm, waste cooking oil and crude Jatropha oil (CJO), and palm fatty acid distillates were successfully converted to biodiesel by the Ti-SBA-15 catalyst at 200 °C. The mesoporous framework gave improved accessibility to the weakly Lewis acidic Ti$^{4+}$ sites, affording higher activity than microporous titanosilicate and TiO$_2$ supports. The Ti-SBA-15 was tolerant of common oil impurities, performing well in the presence of 5 wt% water or 30 wt% FFA. High catalyst loadings of 15 wt% relative to CJO permitted recycling without loss in conversion, although catalyst regeneration between recycles necessitated washing with acetone and subsequent 500 °C calcination.

Most solid acid catalysts employed in biodiesel synthesis are microporous or mesoporous,[32,34,154] properties which the preceding sections highlights are not desirable for accommodating sterically-challenging C$_{16}$–C$_{18}$ TAGs or FFAs for biodiesel synthesis. Incorporation of secondary mesoporosity into a microporous H-β-zeolite to create a hierarchical solid acid significantly accelerated microalgae oil esterification with methanol by lowering diffusion barriers.[155] Templated mesoporous solids are widely used as catalyst supports,[156,157] with SBA-15 silica popular candidates for reactions pertinent to biodiesel synthesis as described above.[142,144,158] However, such surfactant-templated supports possessing long, isolated parallel and narrow channels to not afford efficient in-pore diffusion of bio-oil feedstocks, with resultant poor catalytic turnover. Further improvements in pore architecture are hence required to optimise mass-transport of heavier, bulky TAGs and FFAs common in plant and algal oils. Simulations demonstrate that in the Knudsen diffusion regime,[159] where reactants/products are able to diffuse enter/exit mesopores but experience moderate diffusion limitations, hierarchical pore structures may significantly improve catalyst activity. Materials with interpenetrating, bimodal meso-macropore networks have been prepared using microemulsion[160] or co-surfactant[161]templating routes and are particularly attractive for liquid phase, flow reactors wherein rapid pore diffusion is required.

Liquid crystalline (soft) and colloidal polystyrene nanospheres (hard) templating methods have been combined to create highly organised, macro-mesoporous aluminas[162] and 'SBA-15 like' silicas[163] (Scheme 4), in which both macro- and mesopore diameters can be independently tuned over the range 200–500 nm and 5–20 nm respectively.

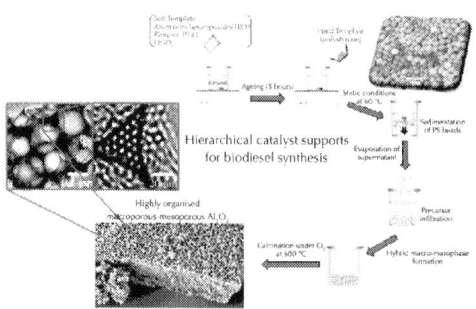

**Scheme 4:** Liquid crystal and polystyrene nanosphere dual surfactant/physical templating route to hierarchical macroporous–mesoporous silicas.

The resulting hierarchical pore network of a propylsulfonic acid functionalised macro-mesoporous SBA-15, illustrates how macropore incorporation confers a striking enhancement in the rates of tricaprylin transesterification and palmitic acid esterification with methanol, attributed to the macropores acting as transport conduits for reactants to rapidly access $PrSO_3H$ active sites located within the mesopores.

ZnO is a heterogeneous photocatalyst which has been used for the degradation of organic pollutants in water and air under UV irradiation[164–167] and for the photoepoxidation of propene by molecular oxygen.[168] $ZnO/SiO_2$ has also been trialled in biodiesel production from crude Mexican *Jatropha curcas* oil *via* a two-step process[169] in which fatty acids were photocatalytically esterified with MeOH under high energy UVC light unrepresentative of the solar spectrum at ground level. Thermally activated transesterification was subsequently performed employing homogeneous NaOH. Porosimetry and IR studies showed no room temperature $CO_2$ or $H_2O$ adsorption suggesting this catalyst should be stable for low temperature esterification. $ZnO/SiO_2$ gave >95% FFA conversion after 8 h of UV irradiation (Fig. 8), with activity constant even after 10 successive runs, although loss of solid catalyst between recycles resulted in a final conversion of only

~20% per run, albeit using very high catalyst loadings. Reaction was proposed to occur *via* FFA adsorption at Lewis acidic $Zn^{2+}$ and MeOH at lattice oxygen, followed by photon adsorption by ZnO and the reaction of photogenerated holes to form $H^+$ and $CH_3O^{\cdot}$ radicals, with photogenerated electrons reacting with adsorbed acids to form $^{\cdot}HOOCR$ radicals; protons and free radicals then reacted to generate intermediates and products. No spectroscopic or chromatographic evidence was presented in support of this elaborate mechanism. Despite the advantages afforded by the $ZnO/SiO_2$ photocatalyst for low temperature FFA esterification, the use of a conventional soluble base in the transesterification step and consequent washing and saponification issues remains problematic, and scale-up of such photocatalysed batch processes to deliver a significant volume of biodiesel will require new photoreactor designs. $ZnO/SiO_2$ materials are also active for the thermally-driven esterification of FFAs (although no details were provided on the nature of these fatty acids) within *Jatropha curcas* crude oils, wherein activity was proportional to acid site density.[170]

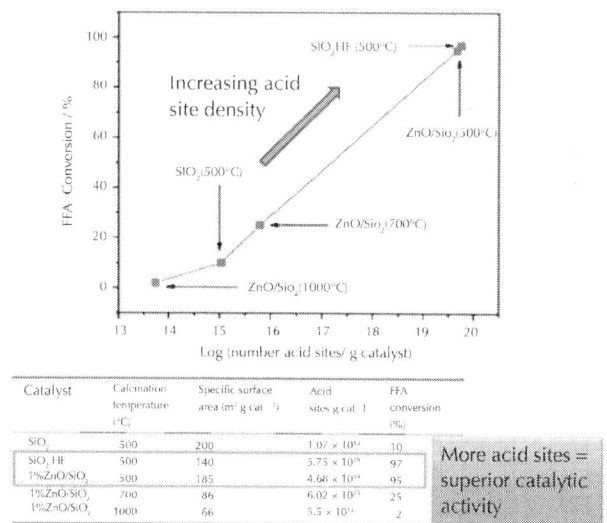

**Figure 8:** Relationship between acid site density and catalytic performance in FFA esterification. Adapted from ref. 171. Copyright (2014), with permission from Elsevier.

In summary, recent developments in tailoring the structure and surface functionality of mesoporous silicas has led to a new generation of tunable solid acid catalysts well-suited to the esterification of short and long chain FFAs, and transesterification of diverse TAGs, with methanol under mild reaction conditions. A remaining challenge is to extend the dimensions and types of pore-interconnectivities present within the host silica frameworks, and to find alternative low cost soft and hard templates to facilitate synthetic scale-up of these catalysts for multi-kg production. Surfactant template extraction is typically achieved *via* energy-intensive solvent reflux, which results in significant volumes of contaminated waste and long processing times, while colloidal templates often require high temperature calcination which prevents template recovery/re-use and releases carbon dioxide. Preliminary steps towards the former have been recently taken, employing room temperature ultrasonication in a small solvent volume to deliver effective extraction of the P123 Pluronic surfactant used in the preparation of SBA-15 in only 5 min, with a 99.9% energy saving and 90% solvent reduction over reflux methods, and without compromising textural, acidic or catalytic properties of the resultant Pr-SO$_3$H-SBA-15 in hexanoic acid esterification (Fig. 9).[171]

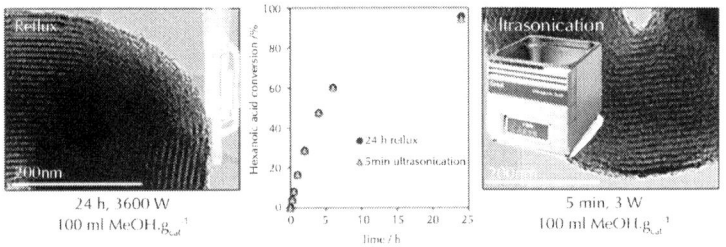

**Figure 9:** Surfactant template extraction *via* energy/atom efficient ultrasonication delivers a one-pot PrSO$_3$H-SBA-15 solid acid catalyst with identical structure and reactivity to that obtained by conventional, inefficient reflux. Adapted from ref. 172 with permission from The Royal Society of Chemistry.

# Heteropolyacids

Heteropolyacids are another interesting class of well-defined acid catalysts, capable of exhibiting superacidity (p$K_H^+$ > 12) and possessing flexible structures.[172] In their native form, heteropolyacids are unsuitable as heterogeneous catalysts for biodiesel applications due to their high solubility in polar media.[173] Dispersing such polyoxometalate clusters over traditional high area oxide supports can modulate their acid site densities,[174,175] but does little to improve their solubility during alcoholysis. Ion-exchanging larger cations into Keggin type phospho- and silicotungstic acids can increase their chemical stability. For example, Cs salts of phosphotungstic acid $Cs_xH_{(3-x)}PW_{12}O_{40}$ and $Cs_yH_{(4-y)}SiW_{12}O_{40}$ are virtually insoluble in water, with proton substitution accompanied by a dramatic increase in surface area of the resulting crystallites.[137,176] As a consequence of these enhanced structural properties, albeit at the expense of losing acidic protons, both $Cs_xH_{(3-x)}PW_{12}O_{40}$ and $Cs_yH_{(y-x)}SiW_{12}O_{40}$ are active for palmitic acid esterification to methyl palmitate and tributyrin transesterification (Fig. 10). For $Cs_xH_{(3-x)}PW_{12}O_{40}$, optimum esterification and transesterification activity was obtained for x= 2.1–2.4, a similar degree of Cs doping to that maximising palmitic acid esterification for $Cs_yH_{(4-y)}SiW_{12}O_{40}$ catalysts (y = 2.8–3.4). These optimal compositions reflect a maximum in the density of accessible surface acid sites within the insoluble Cs-doped catalysts. For $Cs_yH_{(4-y)}SiW_{12}O_{40}$, wherein $C_4$ and $C_8$ TAG transesterification were compared, the absolute reaction rates were faster for the shorter chain triglyceride, attributed to slow in-pore diffusion of the longer chain oil. Absolute TOFs for tributyrin transesterification over the optimised Cs-doped catalyst were greater than for the homogeneous $H_4SiW_{12}O_{40}$ polyoxometalate clusters, a consequence of the greater hydrophobicity of the $Cs_xSiW_{12}O_{40}$ salts compared with the parent $H_4SiW_{12}O_{40}$, which thus afford enhanced activity for the more lipophilic $C_8$ TAG. Optimising the heterogeneous catalytic activity of $Cs_yH_{4-y}SiW_{12}O_{40}$ requires a balance between the retention of acidic protons and generation of stable mesopores to facilitate molecular diffusion. Cs ion-exchange generates interparticle voids large enough to accommodate short-chain TAGs and longer saturated FFAs. Oil/fatty acid and biodiesel polarity and associated mass transport to/from active acid sites is obviously critical in regulating reactivity, and an area where improved materials design in conjunction with molecular dynamics simulations will offer further avenues for high-performance heteropolyacid catalysts.

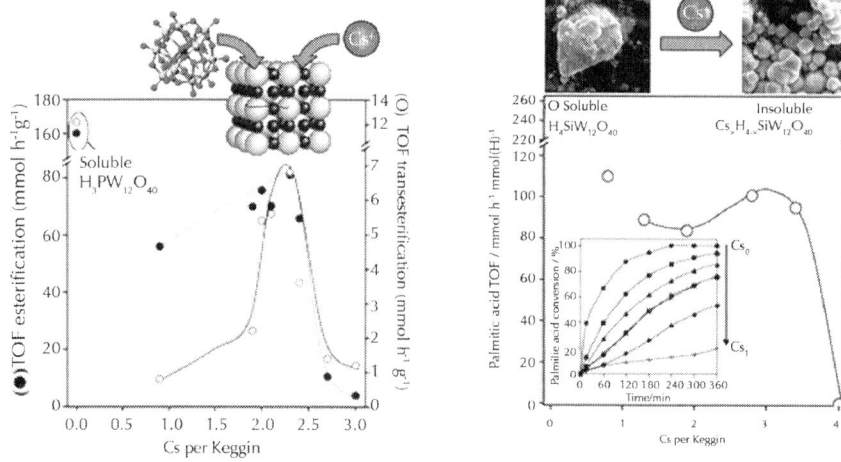

**Figure 10:** Impact of Cs ion-exchange into (left) both $Cs_xH_{(3-x)}PW_{12}O_{40}$ for palmitic acid esterification and tributyrin transesterification with methanol; and (right) and $Cs_yH_{(y-x)}SiW_{12}O_{40}$ for palmitic acid esterification, benchmarked against parent fully protonated, soluble clusters. Adapted from ref. 138 and 177. Copyright (2007 and 2009), wit h permission from Elsevier.

Duan *et al.* have prepared $H_3PW_{12}O_{40}$ supported on magnetic iron oxide particles (MNP-HPA) *via* an acid–base interaction and tested them in palmitic acid esterification with methanol under mild conditions.[177] The magnetic nanoparticles were first coated in a protective $SiO_2$ layer and then functionalised with aminopropyl groups, with the heteropolyacid immobilised by reaction with the amine. Water tolerance was imbued by the addition of nonyl chains to the catalyst surface which lowered the acid loading but improved palmitic acid conversion to 90% at 65 °C. Magnetic separation enabled catalyst recycling without activity loss (Fig. 11), while the presence hydrophobic/oleophilic nonyl groups improved diffusion of the reagent to the active sites, enhancing TOFs compared to the parent MNP-HPA. However, the water tolerance of these materials was limited, with only 1 wt% water reducing FFA conversion to 34%.

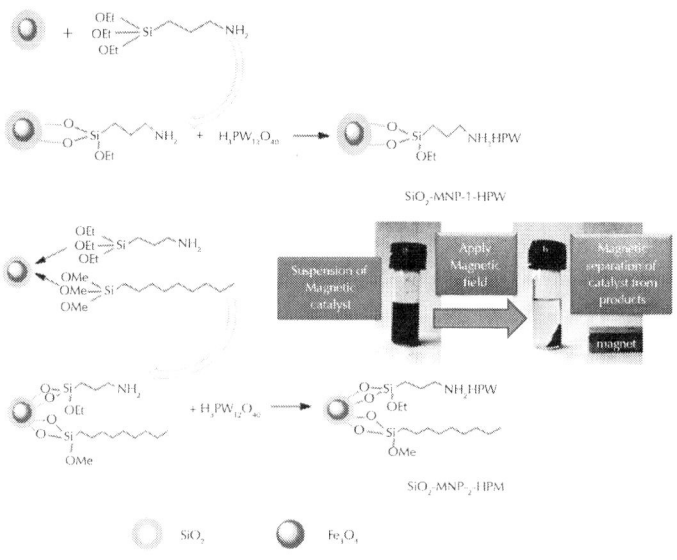

**Figure 11:** Preparation of water-tolerant heteropolyacid on magnetic nanoparticles for palmitic acid esterification. Reprinted from ref. 178 with permission from The Royal Society of Chemistry.

Mesostructured silicas have also been employed as supports for HPAs, for example 12-tungstophosphoric acid (TPA) dispersed over mesoporous MCM-48 is a promising solid acid catalyst for oleic acid esterification with methanol.[178] This catalyst gave 95% conversion to biodiesel with modest alcohol : acid molar ratios, but very high catalyst loadings (30 wt% TPA). Leaching studies employing insensitive colorimetric tests, suggested good catalyst water stability, with minimal loss of W from MCM-48 detectable by atomic absorption (rather than more sensitive ICP), and retention of the majority of acid sites post-reaction (1.50 mmol g$^{-1}$). No explanation was advanced for this extremely surprising water tolerance of TPA, which usually exhibits a high solubility in methanol; entrapment of primary Keggin units within the 3 nm diameter MCM-48 pores seems improbable, and any physical barrier to their dissolution would also likely hinder FFA and FAME access to TPA acid sites. The principal disadvantage of heteropolyacids for esterification and transesterification reactions in short-chain alcohols thus remains their limited water tolerance, which to date can only be overcome through advanced catalyst design and the sacrifice of their high acid strength and site density.

## Acidic polymers and resins

While inorganic frameworks such as SBA-15 or $ZrO_2$ are popular supports for solid acid catalysis, their hydrophilic nature can hinder diffusion of organic reagents. This problem can be avoided by the use of hydrophobic and oleophilic supports, such as mesoporous organic polymers. Sulfonated mesoporous polydivinylbenzene (PDVB) is one such solid acid catalyst,[179] which exhibits absorption capacities for sunflower oil and methanol three times those of $H_3PO_{40}W_{12}$, sulfonated-$ZrO_2$, SBA-15-$SO_3H$ or Amberlyst 15, and consequent superior performance in tripalmitin transesterification, giving an 80% yield of methyl palmitate after 12 h reaction. PDVB-$SO_3H$ proved easily recyclable, with only a modest drop in yield after three recycles, ascribed to a combination of its high surface area, large pore volume, high acid site density, and hydrophobic/oleophilic pore network. Liu et al. utilised an aminophosphonic acid resin based on a polystyrene backbone in the microwave-assisted esterification of stearic acid with EtOH.[180] FAME yields of 90% were obtained after microwave heating to (notionally) 80 °C for 7 h at a catalyst loading of 9 wt%, with slower reaction and a lower limiting conversion of 88% resulting from conventional heating. Kinetic analysis suggested a pseudohomogeneous mechanism in which microwave radiation excited the polar reactants in the solution phase in addition to the solid catalyst. This resin was structurally stable as determined by XRD, TGA and SEM, and recyclable with 87% acid conversion after five uses (Fig. 12).

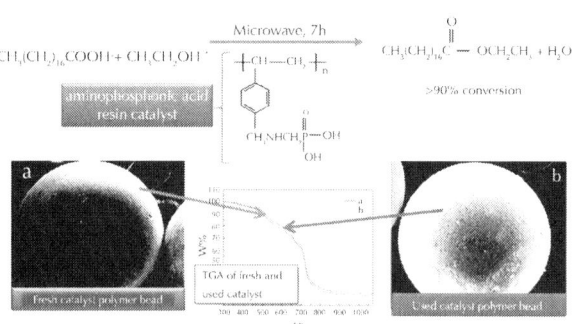

**Figure 12:** Stability of a solid acid resin catalyst for stearic acid esterification. Adapted from ref. 181. Copyright (2013), with permission from Elsevier.

The acid exchange resin, Relite CFS, was tested under batch and continuous modes for the simultaneous esterification and transesterification of oleic acid and soybean oil with methanol,[181]evidencing good activity with 80% FAME obtained after 150 min at 100 °C. Unfortunately this resin was deactivated *via* exchange with metals such as iron present in the feedstream causing catalyst discolouration of beads during continuous operation (Fig. 13); activity could be completely regenerated by suspending the resin in sulphuric acid for 24 h and a further lengthy washing and drying protocol. A copolymer of acidic ionic liquid oligomers and divinylbenzene (PIL) has also been utilised as a catalyst for simultaneous esterification and transesterification of FFA-containing triglyceride mixtures (waste cooking oil), possessing a high acid density of 4.4 mmol g$^{-1}$, high pore volume and surface area of 323 m$^2$ g$^{-1}$, and 35 nm mean pore diameter.[182] The latter and hydrophobic surface character permitted efficient substrate diffusion through the pore network. The PIL copolymer was more active than the acidic ionic liquid alone, giving >99% conversion of oleic acid with MeOH at only 1 wt% catalyst loading. PIL also achieved >99% yield in rapeseed transesterification with MeOH under the same reaction conditions, and proved able to convert high FFA content waste cooking oil into biodiesel with 99% yield in 12 h. The spent catalyst showed no structural changes or loss of acidic sulphur, and hence could be efficiently recycled with almost no loss in performance.

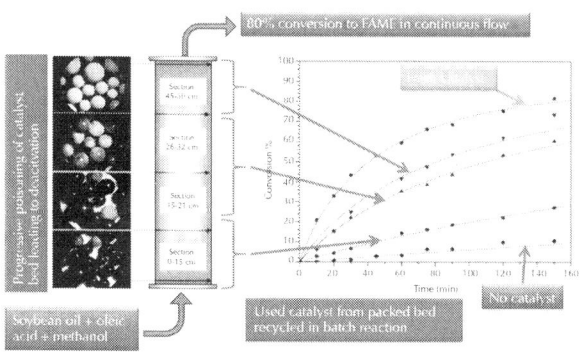

**Figure 13:** Deactivation of an acid resin catalyst during continuous esterification/transesterification of FFA and oil mixtures. Adapted from ref. 182. Copyright (2010), with permission from Elsevier.

# Waste Carbon-Derived Solid Acids

As discussed earlier in this review, many studies have investigated the development of carbon catalysts prepared from second generation biomass such as non-edible crop waste,[2,106,107] algal residues[108]and even waste products from biodiesel production.[109] Sulfonated carbonaceous materials show promising activity for FFA esterification, generally affording higher rates of biodiesel production than commercial resins such Amberlyst with which they are often compared.

Residue of the non-edible seed *Calophyllum inophyllum* has been carbonised to make a biomass-derived solid acid catalyst *via* sulfonation.[107] The resulting catalysts, comprising randomly oriented, amorphous aromatic sheets of low surface area (0.2 to 3.4 $m^2$ $g^{-1}$) and variable acid densities (0.6 to 4.2 mmol $g^{-1}$ dependent on the S wt%), were tested in the simultaneous esterification and transesterification of *Calophyllum inophyllum* seed oil. Esterification activity was greatly proportional to the S loading, but also influenced by the balance of hydrophobic/hydrophilic sites on the carbon which affected diffusion and adsorption of oleo substrates. This balance, and related surface properties, varied with the carbonisation and sulphonation conditions employed; short carbonisation times lead to smaller sheets with higher $SO_3H$ densities and increased activity, but also increased S leaching and concomitant deactivation. Rice husk char was sulfonated with concentrated sulfonic acid under various conditions, and evaluated in the esterification of oleic acid with MeOH.[2] All catalysts were amorphous, with a maximum $SO_3H$ density of 0.7 mmol $g^{-1}$. High conversions were obtained at 110 °C in 2 h for a low alcohol : oil molar ratio of 4 : 1, with the catalyst recyclable and still delivering 84% methyl oleate after seven re-uses despite losing 23% of the initial S through leaching.

Peanut shells processed in a similar manner to that above also yield a strong Brönsted solid acid catalyst, with an acid strength superior to H-ZSM-5 (Si/Al = 75).[183] This catalyst gave >90% conversion of cottonseed oil in methanol transesterification at a methanol : oil molar ratio of only 9 : 1. Recycling and re-use studies employed centrifugation to separate the catalyst, with subsequent acetone washes leading to a 50% reduction in acid site density, although regeneration was achievable by prolonged treatment with 1 M $H_2SO_4$ solution. Despite the environmental compatibility of waste biomass-derived solid acid catalysts, active site retention over prolonged use remains a critical challenge if they are to find implementation in continuous biodiesel production; leaching of sulphate or sulfonic acid groups into the product stream would both shorten catalyst lifetime and degrade fuel quality.

Microalgae are an exciting, potential feedstock for biodiesel production, but following extraction of algal oils, the residue is typically burned or discarded. Fu *et al.*[108] has partially carbonised and sulfonated such residue to create a solid acid catalyst for the esterification of oleic acid and transesterification of triolein with methanol at 80 °C (Fig. 14). Although the resulting catalyst comprised disordered, non-porous aromatic carbon sheets with a very low surface area, the sulfonic acid density of 4.25 mmol $g^{-1}$ afforded an active catalyst with a stable FFA conversion >98% over six sequential oleic acid esterification cycles. The corresponding FAME yield for triolein transesterification was only 22%, but likewise stable across numerous recycles. However, such catalysts were prone to deactivation by adsorbed methanol and hence required regenerative sulphuric acid and hot water washes between recycles. A similar approach was adopted for the waste glycerol by-product of biodiesel production, whereby the polyol was converted *in situ* by partial carbonisation and sulfonation into a solid acid catalyst.[109] High catalyst loadings, reaction temperatures (160 °C) and MeOH : oil ratios (>45) were required to achieve 99% conversion of Karanja oil to FAME, with conversion dropping to only 5% after five recycles, although no analysis of the spent catalyst or leaching studies were reported. Leaching of acid sites was however addressed by Deshmane *et al.*,[184] who investigated sulfonated carbon catalysts prepared from sugar and polyacrylic acid for oleic acid esterification. These catalysts were deactivated by the formation of irregularly-shaped, 1 μm colloidal carbon aggregates, comprised of sulfonated polycyclic hydrocarbons, during the hydrothermal, sulfonation or pulverisation preparative steps, which subsequently leaching into the esterification reaction mixture.

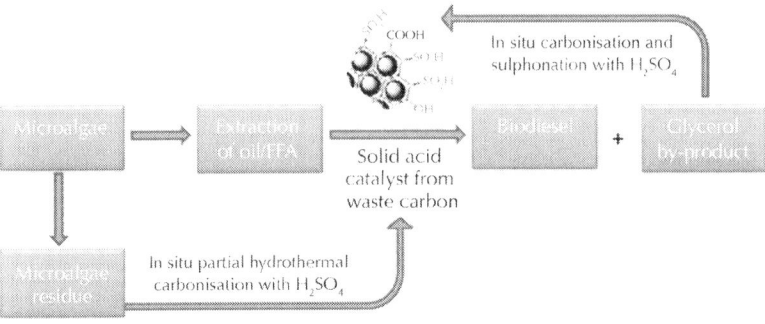

**Fig. 14** Microalgae as a source of bio-oils/fatty acids for biodiesel production, and waste, biomass residue for the synthesis of solid acid catalysts to drive such biodiesel production.

The kinetics of palm oil fatty acid esterification with MeOH over carbonised, sulfonated microcrystalline cellulose (CSMC) have also been compared with those of homogeneous sulphuric acid catalysts,[185] compensating for the phase equilibrium and reaction equilibrium to provide an accurate kinetic reaction model; this approach ensured the biphasic nature of the water–alcohol–oil reaction mixture was correctly represented instead of assuming a pseudo-homogeneous model. Methanol and FFA adsorption over the CSMC was believed a key step in the heterogeneous process, and hence adsorption equilibrium constants were calculated for these molecules along with water and FAME. Unsurprisingly, the free fatty acid was found to adsorb preferentially in the presence of low concentrations of the other molecules. At the start of the esterification reaction, FFA and alcohol were fully miscible, but water and FAME production led to the evolution of two phases; one comprising aqueous methanol and catalyst, and the other methyl ester and unreacted FFA. Mass-transport between these phases is essential, but likely the rate-limiting step. Kinetics of both homogeneously and heterogeneously catalysed biphasic systems were modelled with high conversions favoured by the limited solubility of water in the organic phase, and the use of hydrophobic catalysts which displace water from reaction sites.

A major drawback of the preceding sulfonated carbons is their low surface area, which can be alleviated through the use of carbon nanotubes. Poonjarernsilp and co-workers prepared solid acid catalysts by sulfonating single-walled carbon nanohorns (SWCNHs)[186] which possessed surface areas of 210 m$^2$ g$^{-1}$ and could be further improved by high temperature calcination to open up micropores. The resulting oxidised nanohorns (ox-SWCNs) had surface areas of 1000 m$^2$ g$^{-1}$ and superior pore volumes. However the subsequent sulfonation step required to introduce surface acidity, somewhat lowered the final surface area and pore volume, and drastically altered the pore size distribution, eliminating all the meso- and macropores to leave a narrow range of 2–10 nm pores. Despite the improved morphology of the sulfonated ox-SWCNs relative to the SWCNs, the former had a lower acid site density and was consequently less active in palmitic acid the esterification with methanol; the best yield was obtained for SO$_3$H-SWCNHs, which gave 93% methyl palmitate after 5 h with a catalyst : MeOH : FFA ratio of 0.15 : 0.15 : 5 g. Recycling tests showed a progressive decrease in methyl palmitate yield associated with a loss of acid sites.

## Miscellaneous Solid Acids

A range of additional solid acids have also been investigated, including ferric hydrogen sulphate [Fe(HSO$_4$)$_3$],[187] supported tungsten oxides (WO$_3$/SnO$_2$),[188] supported partially substituted heteropolytungstates,[189] and bifunctional catalysts, such as Mo-Mn/Al$_2$O$_3$-15 wt% MgO,[190] designed to incorporate the benefits of both acid and base catalysis. The iron catalyst had a low surface area of 4–5 m$^2$ g$^{-1}$, and required higher operating temperatures than other solid acids to achieve good biodiesel yields (94% at 205 °C),[187] but was easily recycled by simple washing and drying to remove adsorbed products, maintaining activity over 5 cycles with no evidence of metal leaching. WO$_3$/SnO$_2$ was water tolerant and showed good conversion of soybean oil to FAME at a lower reaction temperature (110 °C), but required high MeOH : oil ratios >30 to achieve a 78% yield,[188] but was prone to on-stream deactivation upon recycling. Tungsten-containing HPAs supported on silica, alumina, and zirconia were also active in biodiesel production from 10 wt% oleic acid in soybean oil delivering FAME yields >75% at a high reaction temperature. Performance was unaffected by the presence of up to 25 wt% of the fatty acid blended with the oil. Cesium addition to the HPA suppressed leaching and thereby improved catalyst stability, resulting in only a 10% fall in biodiesel production after multiple recycles attributed to physical sample loss during product separation. In an attempt to incorporate acid and base character in a single material, Farooq *et al.* prepared a Mo-Mn/γ-Al$_2$O$_3$-15 wt% MgO catalysts *via* wet impregnation of alumina with MgO, followed by impregnation of the γ-Al$_2$O$_3$-MgO with [(NH$_4$)$_6$Mo$_7$O$_{24}$]·4H$_2$O and subsequently aqueous Mn(NO$_3$)$_2$.[190] The resulting thermally processed catalyst possessed highly dispersed MoO$_3$ and MnO acid sites, affording 75% biodiesel yield at 95 °C with a MeOH : oil molar ratio of 15. This bifunctional material could be repeatedly recycled with the yield falling by 20% after 10 uses, a modest deactivation that was attributed to poisoning by strongly adsorbed organics and leaching of the various active metals during transesterification.

# HYDROPHOBICITY STUDIES

The hydrophilic nature of polar silica surfaces hinders their application for reactions involving apolar organic molecules. This is problematic for TAG transesterification (or FFA esterification) due to preferential in-pore diffusion and adsorption of alcohol *versus* fatty acid components. The presence of water in bio-oils (and an inevitable by-product of esterification) can significantly influence biodiesel production, however a major barrier to commercialisation is the development of an efficient, inexpensive and reusable heterogeneous catalyst that can perform at low temperature and pressure.[191] Solid catalysts with ordered and large pores to minimise diffusion limitations, moderate to strong acid sites to overcome the presence of FFAs impurities, and a hydrophobic surface to nullify the effect of water are hence sought.[32,192–196] While solid acid catalysts are of great interest in this regard due to their ability to catalyse both FFA esterification and TAG transesterification,[144,197] sensitivity to water is a common cause of deactivation,[198,199] and water-tolerant solid acids would be highly desirable.[31,37,200] Surface hydrophobicity, and the relative adsorption/desorption rates of reactants/products, are critical parameters influencing (trans)esterification,[201] and tuning catalyst polarity thus offers a route to control competitive adsorption and promote product desorption. Steric factors associated with long fatty acid alkyl chains can also influence reaction rates;[202] Alonso and co-workers explored the relationship between fatty acid polarity/chain length (C$_2$–C$_{16}$) and transesterification rates over solid and liquid acid catalysts.[203] Activity decreased with increasing chain length for a heterogeneous (SAC-13) catalyst, but remained constant when catalysed by $H_2SO_4$, highlighting the negative impact of hydrophilic surfaces on biodiesel production.[203]

Surface hydroxyl groups favour $H_2O$ adsorption, which if formed during FFA esterification can drive the reverse hydrolysis reaction and lowering FAME yields. Surface modification *via* the incorporation of organic functionality into polar oxide surfaces, or dehydroxylation, can lower their polarity and thereby increase initial rates of acid catalysed transformations of liquid phase organic molecules.[204] Surface polarity can also be tuned by incorporating alkyl/aromatic groups directly into the silica framework, for example polysilsesquioxanes can be prepared *via* the co-condensation of 1,4-bis(triethoxysilyl)benzene (BTSB), or 1,2-bis(trimethoxysilyl)-ethane (BTME), with TEOS and MPTS in the sol–gel process[205,206] which enhances small molecule esterification[207] and etherification.[208] This approach has been adopted for the direct synthesis of Lewis acidic, zirconium-containing periodic mesoporous organosilicas (Zr-PMOs), in which zirconocene dichloride was employed as the zirconium source and BTEB was progressively substituted for TEOS.[209] The resulting organosilanes were topologically similar to a purely inorganic Zr-SBA-15 material, but are strongly hydrophobic in nature. Although the one-pot metal doping protocol adopted resulted in relatively low densities of Zr incorporated into the final solid catalyst, hydrophobisation significantly enhanced the per acid site activity in the simultaneous esterification of FFAs and transesterification of TAGs in crude palm oil with methanol at 200 °C, with conversions approaching 90% after only 6 h (Fig. 15). As significant, the catalytic performance of the high organic content Zr-PMO materials was barely influenced by the addition of up to 20 wt% water to the feedstock, in contrast to the inorganic Zr-SBA-15 analogue which was completely poisoned by such water addition. The high water and fatty acid tolerance of these Zr-PMO catalysts renders them especially promising for biodiesel production from waste oil sources.

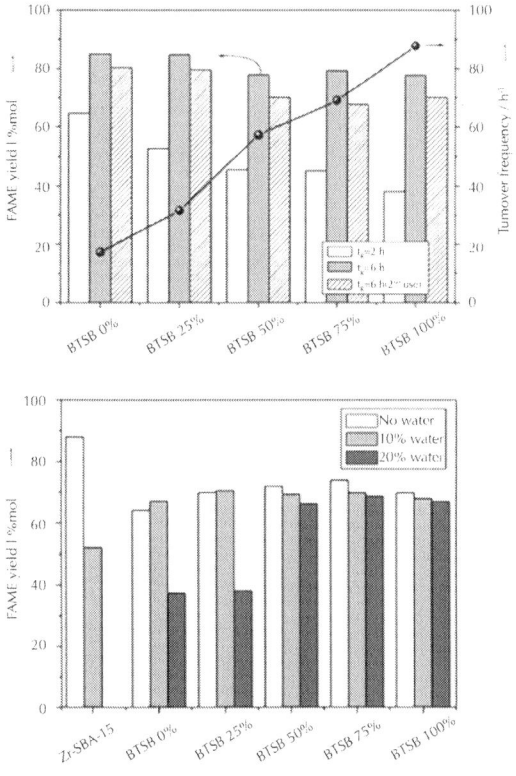

**Figure 15:** (top) FAME yield and turnover frequency calculated for Zr-PMO materials in the methanolysis of crude palm oil highlighting the impact of catalyst hydrophobicity; and (bottom) FAME yield as a function of organic content for Zr-PMO materials in the presence of additional water in the cru de palm oil reaction media evidencing superior water tolerance of hybrid solid acid catalysts. Reprinted from ref. 210. Copyright 2013 John Wiley and Sons.

The incorporation of organic spectator groups (e.g. phenyl, methyl or propyl) during the sol–gel syntheses of SBA-15[210] and MCM-41[211] sulphonic acid silicas is also achievable *via* co-grafting or simple addition of the respective alkyl or aryltrimethoxysilane during co-condensation protocols. An experimental and computational study of sulphonic acid functionalised MCM-41 materials was undertaken in order to evaluate the effect of acid site density and surface hydrophobicity on catalyst acidity and associated performance.[212] MCM-41 was an excellent candidate due to the availability of accurate models for the pore structure from kinetic Monte Carlo simulations,[213] and was modified with surface groups to enable dynamic simulation of sulphonic acid and octyl groups co-attached within the MCM-41 pores. In parallel experiments, two catalyst series were investigated towards acetic acid esterification with butanol (Scheme 5). In one series, the propylsulphonic acid coverage was varied between $\theta$(RSO$_3$H) = 0–100% ML over the bare silica (MCM-SO$_3$H). For the second octyl co-grafted series, both sulfonic acid and octyl coverages were tuned (MCM-Oc-SO$_3$H). These materials allow the effect of lateral interactions between acid head groups and the role of hydrophobic octyl modifiers upon acid strength and activity to be separately probed.

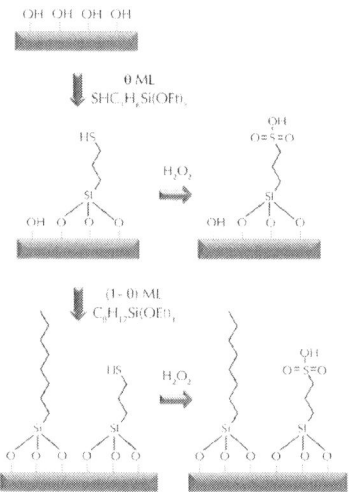

**Scheme 5:** Protocol for the synthesis of sulfonic acid and octyl co-functionalised sulfonic acid MCM-41 catalysts. Adapted from ref. 213 with permission from The Royal Society of Chemistry.

To avoid diffusion limitations, butanol esterification with acetic acid was selected as a model reaction (Fig. 16). Ammonia calorimetry revealed that the acid strength of polar MCM-SO$_3$H materials increases from 87 to 118 kJ mol$^{-1}$ with sulphonic acid loading. Co-grafted octyl groups dramatically enhance the acid strength of MCM-Oc-SO$_3$H for submonolayer SO$_3$H coverages, with $\Delta H_{ads}$(NH$_3$) rising to 103 kJ mol$^{-1}$. The per site activity of the MCM-SO$_3$H series in butanol esterification with acetic acid mirrors their acidity, increasing with SO₃H content. Octyl surface functionalisation promotes esterification for all MCM-Oc-SO₃H catalysts, doubling the turnover frequency of the lowest loading SO₃H material. Molecular dynamic simulations indicate that the interaction of isolated sulphonic acid moieties with surface silanol groups is the primary cause of the lower acidity and activity of submonolayer samples within the MCM-SO₃H series. Lateral interactions with octyl groups help to re-orient sulphonic acid headgroups into the pore interior, thereby enhancing acid strength and associated esterification activity.

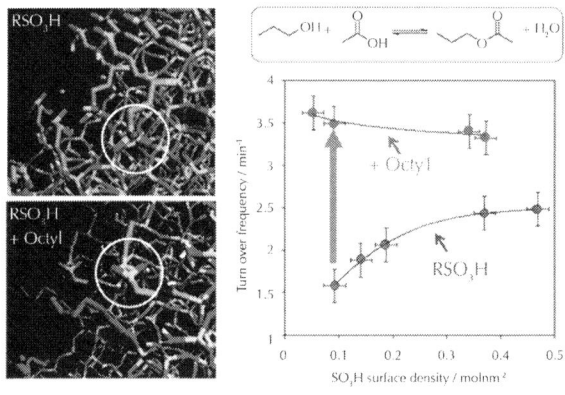

**Figure 16:** (left) Molecular dynamics simulations of MCM-SO₃H and MCM-Oc-SO₃H pore models highlighting the interaction between surface sulfonic acid and hydroxyl groups in the absence of co-grafted octyl chains; (right) influence of PrSO₃H surface density and co-grafted octyl groups on catalytic performance in acetic acid esterification with butanol. Adapted from ref. 213 with permission from The Royal Society of Chemistry.

In some cases, the introduction of hydrophobic functionalities may actually cap the active catalytic site. For example, post-modification of an arene-sulfonic acid SBA-15 by methoxytrimethylsilane deactivated the catalyst by capping the active sites with methyl groups and changing the textural properties, whereas methyl groups introduced *via* a one-pot synthesis did not affect activity towards the microwave-assisted transesterification of soybean oil with 1-butanol.[214] Ethyl groups may also be introduced onto the surface of sulfonic acid modified SBA-15 to impart hydrophobicity. While such ethyl groups has no impact on overall conversions, they improved the initial rate of octanoic acid esterification by displacing reactively-formed water during the start of reaction.[215]

As discussed earlier in this review, hydrophobic solid acid catalysts with large pores are desirable to enhance in-pore mass transport of bulky bio-oils and fatty acids, and to minimise the impact of reactively-formed water during FFA esterification.[37,216] Although many solid catalysts exist with potential in biodiesel production,[154,217] research is increasingly focused on modifying surface hydrophobicity to achieve these goals. Hydrophobicity can be imparted to zeolites by incorporating organic species within their micropores; however, for transesterification involving long chain TAGs, large pore zeolites are preferable, with activity increasing with Si : Al ratio and surface hydrophobicity.[195,218] Fe–Zn double metal cyanides (DMC), possessing only Lewis acid sites, were reported active for sunflower oil transesterification with methanol at 98% conversion. These catalysts exhibited good water tolerance, even in the presence of 20 wt% water in oil, possibly reflecting their surface hydrophobicity and higher coverage of adsorbed reactants.[194] The hydrophobic nature of these catalysts was demonstrated by them in oil–water, water–toluene and water–$CCl_4$ mixtures, wherein the catalyst remained suspended in the hydrophobic layer (Fig. 17).[201,219] Fe–Zn DMC was compared against SZ and Al-MCM-41 for the esterification of long chain ($C_8$–$C_{18}$) FFAs, and the transesterification of soybean oil. SZ and Al-MCM-41 showed better conversion than DMC towards the fatty acids, but reverse was observed for the more hydrophobic soybean oil.[201] Fe–Zn DMC possessed a hybrid structure containing both crystalline and amorphous phases; hydrophobicity ascribed to the presence of the latter phase.[220]

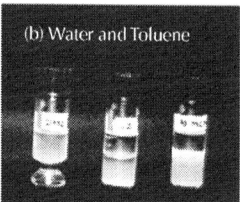

**Figure 17:** Preferential dispersion of DMC in the nonpolar, organic phase, and SZ and Al-MCM-41 in the polar aqueous phase of (a) water–$CCl_4$ and (b) water–toluene solvent mixtures. Reprinted with permission from ref. 202. Copyright 2010 American Chemical Society.

Cesium-doped dodecatungstophosphoric acid (CsPW) has shown promise as a water-tolerant solid acid catalyst for the hydrolysis of ethyl acetate,[221] and found subsequent employ in the transesterification of Eruca sativa Gars (ESG) oil.[202] The authors claimed that CsPW exhibited excellent water-tolerance towards ESG transesterification, despite oil conversions falling by ~90% upon the addition of only 1% water. Zn containing HPAs display more impressive credentials for transforming challenging feedstocks, with zinc dodecatungstophosphate nanotubes possessing Lewis and Brönsted acid sites effective for the for the simultaneous esterification and transesterification of palmitic acid, and transesterification of waste cooking oils with 26% FFA and 1% water.

The one-pot synthesis of a styrene modified sulfonic acid silica 15 was achieved by adding styrylethyl-trimethoxysilane during a conventional SBA-15 synthesis.[222] Styryl groups polymerised on the silica surface imparted hydrophobicity. Subsequent acid functionalisation of these materials resulted in a polystyrene-modified sulfonic acid SBA-15, which was active for oleic acid esterification with $n$-butanol, and proved superior to SAC-13 and Amberlyst-15 due to the hydrophobic polystyrene coating and high surface area.[223]

Surface acidity has also been imparted to hydrophobic, mesoporous polydivinylbenzene (PDVB) by sulfonic acid grafting. Such materials were employed in tripalmitin transesterification with methanol, revealing that mesoporous PDVB with electron withdrawing $-SO_3H-SO_2CF_3$ groups gave good activity with 91% yield maintained up to 5 re-uses. Contact angle measurements confirmed the hydrophobic nature and high oleophilicity of these materials. PDVB grafted with chlorosulfonic acid also generated hydrophobic solid acid catalysts for tripalmitin which were successfully transesterification whose performance (80% methyl palmitate yield) was superior to HPA, SBA-15-$SO_3H$, Amberlyst 15, and mesoporous $SO_4-ZrO_2$. The same activity trend was observed for sunflower oil transesterification wherein all $C_{16}-C_{27}$ fatty acids were converted to FAMEs reflecting the higher adsorption capacity and hence reactivity of these PDVB acids.[179,223,224] Polyaniline functionalised with methanosulfonic (MSA-Pani), camphorosulfonic (CSA-Pani) and lignosulfonic (LG-Pani) acids and polyaniline sulfate (S-Pani) also show promise in biodiesel synthesis with the LG-Pani catalyst possessing the greatest acid site density (3.62 $mmol_H^+ g^{-1}$) and highest conversion due to the close proximity of hydrophobic centres to the active sites. Sulfonic acid containing ionic liquids have also been co-polymerised with divinyl benzene, to form a hydrophobic, solid acidic ionic liquid polymer (PIL) for the transesterification of rapeseed and waste cooking oils, outperforming homogeneous counterparts.[182]

Partial carbonisation and sulfonation of organic matter offers a route to combine acidity and hydrophobicity into carbon based mesoporous materials.[225,226] Such solids are typically partially amorphous, but offer efficient transesterification of non-edible seed oils.[107] It has proven difficult to introduce organic groups into the surface of ordered mesoporous carbons (OMCs) prepared through high temperature carbonisation, however surface pretreatment with $H_2O_2$ to introduce hydroxyl anchors enables their subsequent sulfonation and a resulting hydrophobic and stable acid catalyst for oleic acid esterification.[227] Sulfonated single-walled carbon nanotubes ($SO_3H$-SWCH) have also been investigated for palmitic acid esterification, exhibiting higher activity than other sulfonated carbons, such as oxidized SWCNHs (ox-SWCNHs), activated carbon (AC), and carbon black (CB), attributed to the stronger acidity of $SO_3H$-SWCH and hydrophobicity of the carbon surface in the vicinity of acid sites,[186] enabling it to even outperform liquid $H_2SO_4$. Another interesting class of porous hydrophobic catalysts are mesoporous titanosilicates which are active for biodiesel and biolubricant synthesis. Ti incorporation into the surface of mesoporous SBA-12 and SBA-16 generates Lewis acid sites which are active for esterification and transesterification. The high activity of these Lewis acid sites is comparable to that observed for Fe–Zn double metal cyanides.[194] Solid state $^{29}Si$ NMR studies show that Ti-SBA-16 is more hydrophobic than Ti-SBA-12. In biolubricant synthesis, for which surface hydrophobicity is crucial, Ti-SBA-16 is significantly more active than Ti-SBA-12.[228]

Lipase has also been immobilised on hydrophobic supports with a view to transesterifying water containing oils,[229] wherein small amounts of water improved lipase activity.[230] The application of lipase enzymes can be made more cost-effective by heterogenisation over a solid support, with hydrophobic supports both assisting lipase surface attachment and promoting FFA esterification and bio-oil transesterification. *Burkholderia* lipase supported on hydrophobic magnetic particles for olive oil transesterification gave 70% conversion to FAME even in the presence of up to 10% water and was readily recycled.[231] FAME production from canola oil was also achieved using lipase immobilised on a hydrophobic, microporous styrene-divinylbenzene copolymer, wherein the support hydrophobicity mitigated the inhibitory effect of water and glycerol affording a 97% yield.[232]

Solid basic hydrotalcites also showed enhanced activity and reusability for soybean oil transesterification when dispersed over polyvinylalcohol (PVA) membranes, although increasing the hydrophobicity *via* polymer cross-linking lowered activity, presumably due to poor active site accessibility by the bulky substrate. Hydrophilicity *versus* hydrophobicity may be tuned over such membranes by succinic anhydride and acetic anhydride treatments, with a mix of hydrophilic and hydrophobic environments near the active hydrotalcite sites required for optimal transesterification.[233]An interesting contrast to the preceding systems (wherein water poisons FAME formation) was reported for CaO catalysed soybean transesterification, for which small amounts of water actually improve activity, attributed to an increase in the concentration of surface OH- active base sites.[234] Mixed MgO–CaO also exhibited a surprising water tolerance in rapeseed oil transesterification, enabling 98% conversion with 2% water, with $La_2O_3$–CaO active even in the presence of 10% water.[235,236]

Periodic Mesoporous Organosilicas (PMOs) are a promising class of materials that can be used as catalyst supports for biodiesel production. PMOs are hybrid organic–inorganic materials with mesopore networks akin to SBA-15.[236] Functionalisation of PMOs with catalytically active organic moieties is an emergent field of heterogeneous catalysis, and since the organic groups are dispersed throughout the framework (rather than confined to hydroxylated patches of the surface[212]), active sites and hydrophobic centres can be co-located in high concentrations. Methylpropyl sulfonic acid functionalised phenylene- and ethyl-bridged PMOs have been synthesised and tested for the transesterification of sunflower oil, canola oil, corn oil, refined olive oil and olive sludge.[237] These functionalised PMOs gave comparable or better activity than SBA-15-PrSO$_3$H under optimised conditions, with the ethyl-bridged PMO showing highest activity with a 98% yield. Water adsorption studies proved that the phenylene-bridged PMO was more hydrophobic than the ethyl-bridged variant, but less active, showing that a balance of hydrophobic *versus* hydrophilic mesostructural properties are necessary for optimum transesterification. Heterogeneous catalysts with tunable hydrophobicity, acid/base character, and good thermal stability, whether based upon polymeric or inorganic frameworks, are hence promising new solutions to TAG transesterification and FFA esterification of high moisture content feedstocks.

# INFLUENCE OF REACTOR DESIGN AND OPERATING CONDITIONS

One other development likely to impact on the commercial exploitation of heterogeneous catalysts for biodiesel production is the design of innovative chemical reactors to facilitate continuous processing of viscous bio-oils. Although many industrial biodiesel production plants operate in batch mode at a significant scale ($\sim$7000 tons year$^{-1}$),[238–240] there is a need to move towards heterogeneously catalysed, continuous flow reactors in order to avoid the separation issues of homogeneous catalysts and drawbacks of batch mode (notably increased capital investment required to run at large volumes and increased labour costs of a start/stop process)[241] and increase the scale of operation (8000–125 000 tons year$^{-1}$).[239,240] A range of process engineering solutions have been considered for the continuous esterification of FFAs, including the use of fixed bed[242] or microchannel-flow reactors,[243] pervaporation methods,[244] and reactive distillation.[245,246] Process intensification methods in biodiesel production have been reviewed in depth elsewhere.[247,248]

Reactive distillation combines chemical conversion and separation steps in a single stage. This simplifies the process flow sheets, reduces production costs, and extends catalyst lifetimes through the continuous removal of water from the system. However, this technique is only applicable if the reaction is compatible with the temperatures and pressures required for the distillation. Kiss *et al.*demonstrated this approach for the esterification of dodecanoic acid with a range of alcohols catalysed by sulphated zirconia.[245] Their reactive distillation was 100% selective, permitted shorter residence times than comparable flow systems, and did not require excess alcohol. The latter is a major advantage over the overwhelming majority of conventional biodiesel syntheses wherein, since reaction between the triglyceride and alcohol is reversible, large alcohol excesses are normally required to achieve full conversion (the excess alcohol must then be separated and re-used to ensure economic process viability).

Any continuous flow reactor must be designed appropriately to harness the full potential of the integrated heterogeneous catalyst; plug flow is a desirable characteristic since it permits tight control over the product composition, and hence minimises downstream separation processes, and associated capital investment and running costs. Conventional plug flow reactors are ill-suited to slow reactions such as FFA esterification and TAG transesterification, since they require very high length : diameter ratios to achieve good mixing, and in any event are problematic due to their large footprints and pumping duties, and control difficulties. Oscillatory Baffled Reactors (OBRs) circumvent these problems by oscillating the reaction fluid through orifice plate baffles to achieve efficient mixing and plug flow,[249] thereby decoupling mixing from the net fluid flow in a scalable fashion, enabling long reaction times on an industrial scale, and have been applied to homogeneously catalysed biodiesel synthesis.[250] Vortical mixing in the OBR also offers an effective, controllable method of uniformly suspending solid particles and was recently utilised to entrain a $PrSO_3H$-SBA-15 mesoporous silica within a glass OBR under an oscillatory flow for the continuous esterification of propanoic, hexanoic, lauric and palmitic acid (Fig. 18).[42] Excellent semi-quantitative agreement was obtained between the kinetics of hexanoic acid esterification within the OBR and a conventional stirred batch reactor, with fatty acid chain length identified as a key predictor of solid acid activity. Continuous esterification within the OBR improved ester yields compared with batch operation due to water by-product removal from the catalyst reaction zone, evidencing the versatility of the OBR for heterogeneous flow chemistry and potential role as a new clean catalytic technology.

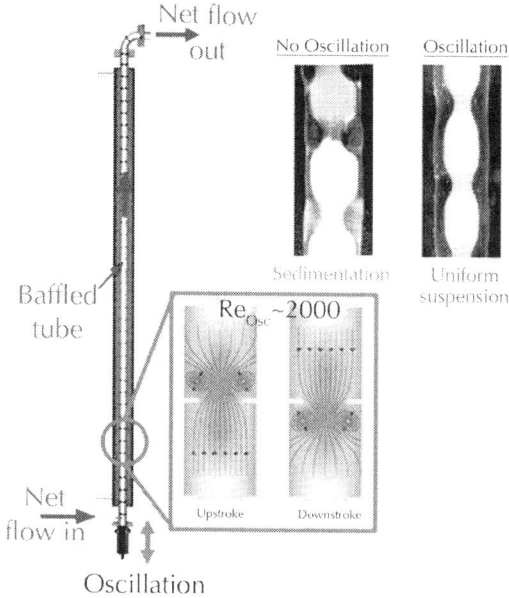

**Figure 18:** Schematic of reactor flow and mixing characteristics within an OBR, and associated optical images of a $PrSO_3H$-SBA-15 solid acid powder without oscillation (undergoing sedimentation) or with a 4.5 Hz oscillation (entrained within baffles). Adapted from ref. 42 with permission from The Royal Society of Chemistry.

Phase equilibria considerations are very important in biodiesel production *via* TAG transesterification with methanol, since the reactant and alcohol are generally immiscible, whereas the FAME product is miscible, hampering mass transport and retarding reaction. Separation and purification of the product phase, a mixture of solid catalyst, unreacted oil, glycerol and biodiesel, adds further complexity and cost to production.[251] These problems may be alleviated through the use of membrane reactors,[252-256] wherein the reactor walls are made of a semi-permeable material designed to allow passage of the FAME/glycerol phase, while retaining the oil-rich/MeOH emulsion for further reaction. Xu *et al.* utilised a MCM-41 supported *p*-toluenesulfonic acid catalyst to pack a ceramic membrane tube for the transesterification of a recirculating soybean oil and methanol feed (Fig. 19a). A higher biodiesel yield was obtained with the membrane reactor than with a homogeneous *p*-toluenesulfonic acid catalyst under comparable conditions in batch mode (84 *versus* 66%). Catalyst re-used evidenced only a minor loss of activity (92% of original after the third cycle).[254] Biodiesel yield was a strong function of circulation velocity; low velocities improved permeation efficiency, while high velocities enhanced reactant mixing intensity. Although membrane reactors offer efficient transesterification and separation, they require high catalyst volumes, for example a 202 cm$^3$ continuous reactor employed 157 g of a microporous TiO$_2$/Al$_2$O$_3$ membrane packed with potassium hydroxide supported on palm shell activated carbon to produce high quality methyl esters from palm oil (Fig. 19b).[252]

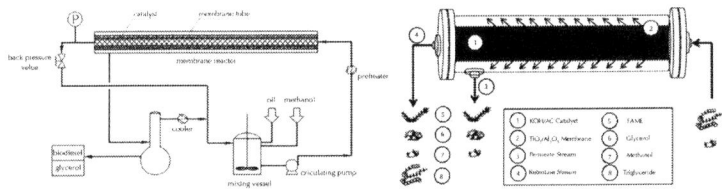

**Figure 19:** Schematic of recirculating packed membrane reactors for continuous biodiesel production*via* (a) solid acid and (b) base catalysts. Reprinted from ref. 252 and 254. Copyright (2011 and 2014), with permission from Elsevier.

Enzymatic catalysed biodiesel production has been reported in both continuous[257,258] and batch modes.[259] Nature has developed a range of lipase biocatalysts for the selective synthesis of FAME at low reaction temperature, which are tolerate to high FFA levels.[260,261] Immobilisation on solid supports enables such biocatalysts to be used in continuous mode with low methanol : oil ratios.[262]However, there are numerous shortcomings of biocatalysts including high enzyme costs, long residence times, and low biodiesel yields. Some enzymes can also be deactivated by short chain alcohols and the glycerol by-product;[263] this problem can be overcome through the use of organic solvents to extract the alcohols and glycerol, but this adds further complexity and cost, and weakens the green credentials of biodiesel production. Enzymes must also operate in the presence of water in order to avoid denaturation, however this additional water must be subsequently removed from the resulting fuel to meet biodiesel standards (<0.05 vol% H$_2$O), these drying steps introducing further costs. An alternative approach is the use of near-critical[264] or supercritical CO$_2$[255,256] as a reaction medium to minimise enzyme inhibition by methanol, enhance oil solubility and diffusion, and assist catalyst/biodiesel separation *via* simple depressurisation. The associated strengths and weaknesses of supercritical biodiesel production are reviewed elsewhere.[265]

Ultrasound[266,267] and microwaves[268,269] have been explored as a means of eliminating heat and mass transfer limitations, and shortening residence times to achieve high biodiesel conversions. Ultrasound was used by Gude *et al.* in place of thermal heating for the transesterification of waste cooking oil,[266] allowing efficient heating to a temperature of 60–65 °C and lowering reaction times to 1–2 min. Chand *et al.* observed similar improvements in heat transfer and reaction time applying ultrasonication to soybean oil transesterification.[270] However, both groups employed a homogeneous NaOH catalyst, hindering product purification. Ultrasound was used with a heterogeneous catalyst for continuous biodiesel production from palm oil by Salamatinia *et al.*[271] BaO and SrO catalysts were tested, and ultrasound again found to reduce the reaction times and catalyst loadings needed to achieve >95% FAME yields. Cost analysis of an ultrasonic process suggests it would be at least three times more expensive to run than a conventionally heated continuous biodiesel reactor.[270] The origin of ultrasonic enhancements in respect of reaction mixing *via e.g.* cavitation or micro-streaming, remains a matter of debate.[272] Microwaves have been coupled with continuous flow reactors for the transesterification of waste cooking oil, accelerating biodiesel production compared to conventional thermal heating, and hence higher throughput.[269] The majority of microwave studies to date have focused on homogeneously catalysed processes, although some innovative combinations of waste derived (eggshell) solid catalysts and microwaves are emerging.[273] Such microwave systems also require less solvent and catalyst. However, microwave penetration depth is a limiting factor[268] which may restrict scale-up from laboratory reactor designs, and uncontrolled and irregular heat distribution can result in 'hot spots' and 'cold spots'.[267,268]

# FUTURE DIRECTIONS

If sourced and produced in a sustainable fashion, biodiesel has the potential to play an important role in meeting renewable fuel targets. However, developments in materials design and construction are critical to achieve significant improvements in heterogeneously catalysed biodiesel production. Designer solid acid and base catalysts with tailored surface properties and pore networks offer process improvements over existing, commercial homogeneous catalysed production employing liquid bases, facilitating simple catalyst separation and fuel purification, coupled with continuous biodiesel synthesis. Tuning the surface hydrophobicity of heterogeneous catalysts can strongly influence oil transesterification and FFA esterification through the expulsion of water away from active catalytic centres, thus limiting undesired reverse hydrolysis processes, notably in high water content waste oils. Solid materials capable of simultaneous FFA esterification and TAG transesterification under mild conditions present a major challenge for catalytic scientists, although (insoluble) high area superacids represent a step in this direction. We predict that in the future, hierarchical solid acids may be employed to first hydrolyse non-edible oil feedstocks, and subsequently esterify the resulting FFAs to FAME. Synthesis of nanostructured (e.g. nanocrystalline) catalysts and the application of surface-initiated, controlled polymerisation to functionalise oxide surfaces with polymeric organic species to create hybrid organic–inorganic architectures with high active site loadings, will prove valuable in the quest for enhanced catalyst performance.

Despite concerns over long term biodiesel use in high performance engines, the implementation of FAME containing longer chain ($>C_{18}$) esters in heavy-duty diesel engines should prove less problematic to on short timecales. However, the widespread uptake and development of next-generation biodiesel fuels requires progressive government policies and incentive schemes to place biodiesel on a comparative footing with (heavily subsidised) fossil-fuels. Blending of biodiesel with pyrolysis oil derived from lignocellulosic waste is an attractive route to power low-medium scale Combined Heat and Power (CHP) engines. Increasing use of waste or low grade oil sources remains a challenge for existing heterogeneous catalysts, since the high concentration of impurities (acid, moisture, heavy metals) induce rapid on-stream deactivation, and necessitate improved upstream oil purification, or more robust catalyst formulations tolerant to such components. Feedstock selection is dominated by regional availability, however the drive to use non-edible oil sources in areas where they cannot be readily sourced will require close attention to the entire supply chain and emissions/costs associated with new transportation networks, and may favour genetic modification of plant and algal strains to adapt to non-native climates.

The viscosity and attendant poor miscibility of many oil feedstocks with light alcohols continues to hamper the use of new heterogeneous catalysts for continuous biodiesel production, from both a materials and engineering perspective. Future process optimisation and growth in biodiesel supply and demand needs a concerted effort between catalyst chemists, chemical engineers and experts in molecular simulation in order to take advantage of innovative reactor designs and develop catalysts and reactors in tandem. Alternative reactor technologies and process intensification *via e.g.* reactive distillation and oscillatory flow reactors will facilitate distributed biodiesel production. It is essential that technical advances in both materials chemistry and reactor engineering are pursued if biodiesel is to remain a key player in the renewable energy sector during the 21st century.

# ACKNOWLEDGEMENTS

A.F.L. thanks the EPSRC for the award of a Leadership Fellowship (EP/G007594/4). K.W. thanks the Royal Society for the award of an Industry Fellowship.

# REFERENCES

1.   S. Kretzmann, http://priceofoil.org/2013/11/26/new-analysis-shows-growing-fossil-reserves-shrinking-carbon-budget/.

2.   C. C. Authority, Reducing Australia's Greenhouse Gas Emissions – Targets and Progress Review Draft Report, Commonwealth of Australia, 2013.

3.   C. C. Secretariat, The critical decade 2013 Climate change science, risks and responses, Commonwealth of Australia, 2013.

4.   I. E. Agency, *Prospect of limiting the global increase in temperature to 2 °C is getting bleaker*, http://www.iea.org/newsroomandevents/news/2011/may/name,19839,en.html.

5.   I. E. Agency, Redrawing the Energy Climate Map, 2013.

6.  U. S. E. I. Administration, *International Energy Outlook 2013*, 2013.

7.  PwC, World in 2050. The BRICs and beyond: prospects, challenges and opportunities, 2013.

8.  N. Armaroli and V. Balzani, *Angew. Chem., Int. Ed.*, 2007, 46, 52–66.

9.  P. Azadi, O. R. Inderwildi, R. Farnood and D. A. King, *Renewable Sustainable Energy Rev.*, 2013, 21, 506–523

10. J. M. Thomas, *Proceedings of the Royal Society A: Mathematical, Physical and Engineering Science*, 2012, 468, 1884–1903.

11. G. A. Somorjai, H. Frei and J. Y. Park, *J. Am. Chem. Soc.*, 2009, 131, 16589–16605.

12. A. Demirbas, *Energy Policy*, 2007, 35, 4661–4670 CrossRef PubMed.

13. T. P. Vispute, H. Zhang, A. Sanna, R. Xiao and G. W. Huber, *Science*, 2010, 330, 1222–1227.

14. P. M. Mortensen, J. D. Grunwaldt, P. A. Jensen, K. G. Knudsen and A. D. Jensen, *Appl. Catal., A*, 2011, 407, 1–19.

15. R. Luque, L. Herrero-Davila, J. M. Campelo, J. H. Clark, J. M. Hidalgo, D. Luna, J. M. Marinas and A. A. Romero, *Energy Environ. Sci.*, 2008, 1, 542–564

16. C. S. K. Lin, L. A. Pfaltzgraff, L. Herrero-Davila, E. B. Mubofu, S. Abderrahim, J. H. Clark, A. A. Koutinas, N. Kopsahelis, K. Stamatelatou, F. Dickson, S. Thankappan, Z. Mohamed, R. Brocklesby and R. Luque, *Energy Environ. Sci.*, 2013, 6, 426–464

17. D. W. McLaughlin, *Conserv. Biol.*, 2011, 25, 1117–1120.

18. F. Danielsen, H. Beukema, N. D. Burgess, F. Parish, C. A. Brühl, P. F. Donald, D. Murdiyarso, B. E. N. Phalan, L. Reijnders, M. Struebig and E. B. Fitzherbert, *Conserv. Biol.*, 2009, 23, 348–358.

19. W. M. J. Achten, L. Verchot, Y. J. Franken, E. Mathijs, V. P. Singh, R. Aerts and B. Muys, *Biomass Bioenergy*, 2008, 32, 1063–1084.

20. T. M. Mata, A. A. Martins and N. S. Caetano, *Renewable Sustainable Energy Rev.*, 2010, 14, 217–232.

21. BP, *BP Energy Outlook 2030*, 2011.

22. G. Knothe, *Top. Catal.*, 2010, 53, 714–720.

23. M. J. Climent, A. Corma, S. Iborra and A. Velty, *J. Catal.*, 2004, 221, 474–482.

24. U. Constantino, F. Marmottini, M. Nocchetti and R. Vivani, *Eur. J. Inorg. Chem.*, 1998, 1439–1446.

25. K. Narasimharao, A. Lee and K. Wilson, *J. Biobased Mater. Bioenergy*, 2007, 1, 19–30.

26. M. R. Othman, Z. Helwani, Martunus and W. J. N. Fernando, *Appl. Organomet. Chem.*, 2009, 23, 335–346.

27. Y. Liu, E. Lotero, J. G. Goodwin and X. Mo, *Appl. Catal., A*, 2007, 33, 138–148.

28. J. Geuens, J. M. Kremsner, B. A. Nebel, S. Schober, R. A. Dommisse, M. Mittelbach, S. Tavernier, C. O. Kappe and B. U. W. Maes, *Energy Fuels*, 2007, 22, 643–645.

29. J. Hu, Z. Du, Z. Tang and E. Min, *Ind. Eng. Chem. Res.*, 2004, 43, 7928–7931.

30. G. Knothe, *Fuel Process. Technol.*, 2005, 86, 1059–1070.

31. F. Ma and M. A. Hanna, *Bioresour. Technol.*, 1999, 70, 1–15

32. E. Lotero, Y. Liu, D. E. Lopez, K. Suwannakarn, D. A. Bruce and J. G. Goodwin, *Ind. Eng. Chem. Res.*, 2005, 44, 5353–5363.

33. R. Luque, J. C. Lovett, B. Datta, J. Clancy, J. M. Campelo and A. A. Romero, *Energy Environ. Sci.*, 2010, 3, 1706–1721.

34. J.-P. Dacquin, A. F. Lee and K. Wilson, *Thermochemical Conversion of Biomass to Liquid Fuels and Chemicals*, The Royal Society of Chemistry, 2010, pp. 416–434.

35. K. Wilson and A. F. Lee, *Catal. Sci. Technol.*, 2012, 2, 884.

36. L. J. Konwar, J. Boro and D. Deka, *Renewable Sustainable Energy Rev.*, 2014, 29, 546–564.

37. A. Islam, Y. H. Taufiq-Yap, C.-M. Chu, E.-S. Chan and P. Ravindra, *Process Saf. Environ. Prot.*, 2013, 91, 131–144.

38. K. Ramachandran, T. Suganya, N. Nagendra Gandhi and S. Renganathan, *Renewable Sustainable Energy Rev.*, 2013, 22, 410–418.

39. Y. M. Sani, W. M. A. W. Daud and A. R. Abdul Aziz, *Appl. Catal., A*, 2014, 470, 140–161.

40. I. M. Atadashi, M. K. Aroua, A. R. Abdul Aziz and N. M. N. Sulaiman, *J. Ind. Eng. Chem.*, 2013, 19, 14–26.

41.    Y. M. Sani, W. M. A. W. Daud and A. R. Abdul Aziz, *J. Environ. Chem. Eng.*, 2013, 1, 113–121.

42.    V. C. Eze, A. N. Phan, C. Pirez, A. P. Harvey, A. F. Lee and K. Wilson, *Catal. Sci. Technol.*, 2013, 3, 2373–2379.

43.    S. P. Singh and D. Singh, *Renewable Sustainable Energy Rev.*, 2010, 14, 200–216.

44.    P. Schenk, S. Thomas-Hall, E. Stephens, U. Marx, J. Mussgnug, C. Posten, O. Kruse and B. Hankamer, *BioEnergy Res.*, 2008, 1, 20–43.

45.    R. E. H. Sims, W. Mabee, J. N. Saddler and M. Taylor, *Bioresour. Technol.*, 2010, 101, 1570–1580.

46.    J. Kansedo, K. T. Lee and S. Bhatia, *Biomass Bioenergy*, 2009, 33, 271–276.

47.    J. M. Encinar, J. F. Gonzalez, A. Pardal and G. Martinez, *Fuel Process. Technol.*, 2010, 91, 1530–1536.

48.    J. Calero, D. Luna, E. D. Sancho, C. Luna, F. M. Bautista, A. A. Romero, A. Posadillo and C. Verdugo, *Fuel*, 2014, 122, 94–102 CrossRef CAS PubMed.

49.    V. Scholz and J. N. da Silva, *Biomass Bioenergy*, 2008, 32, 95–100

50.    E. Akbar, Z. Yaakob, S. K. Kamarudin, M. Ismail and J. Salimon, *Eur. J. Sci. Res.*, 2009, 29, 396–403.

51.    A. Karmakar, S. Karmakar and S. Mukherjee, *Renewable Sustainable Energy Rev.*, 2012, 16, 1050–1060.

52.    Y. Chisti, *Biotechnol. Adv.*, 2007, 25, 294–306 CrossRef CAS PubMed.

53.    H. N. Bhatti, M. A. Hanif, M. Qasim and R. Ataur, *Fuel*, 2008, 87, 2961–2966.

54.    D. M. Kargbo, *Energy Fuels*, 2010, 24, 2791–2794.

55.    D. Frieden, N. Pena, D. N. Bird, H. Schwaiger and L. Canella, *Center for International Forestry Research (CIFOR)*, Bogor, Indonesia, 2011,

56.    L. Lardon, A. Hélias, B. Sialve, J.-P. Steyer and O. Bernard, *Environ. Sci. Technol.*, 2009, 43, 6475–6481.

57.    ASTM Standard D6751, *Standard Specification for Biodiesel Fuel Blend Stock (B100) for Middle Distillate Fuels*, ASTM

International, West Conshohocken, PA, 2012, DOI: 0.1520/ D6751-12, www.astm.org.

58. ASTM Standard D6751, *Standard Specification for Biodiesel Fuel Blend Stock (B100) for Middle Distillate Fuels*, ASTM International, West Conshohocken, PA, 2012, DOI: 0.1520/ D6751-12, www.astm.org.

59. D. Y. C. Leung, X. Wu and M. K. H. Leung, *Appl. Energy*, 2010, 87, 1083–1095.

60. Y. C. Sharma and B. Singh, *Renewable Sustainable Energy Rev.*, 2009, 13, 1646–1651.

61. A. Demirbas, *Energy Convers. Manage.*, 2009, 50, 14–34.

62. A. Elbehri, A. Segerstedt and P. Liu, *Biodiesel and sustainability challenge: A global assessment of sustainability issues, trends and policies for biofuels and related feedstocks*, Food and Agriculture Organization of the United Nations, 2013

63. R. Sarin, M. Sharma, S. Sinharay and R. K. Malhotra, *Fuel*, 2007, 86, 1365–1371.

64. M. K. Lam, K. T. Lee and A. R. Mohamed, *Biotechnol. Adv.*, 2010, 28, 500–518.

65. J. C. Escobar, E. S. Lora, O. J. Venturini, E. E. Yáñez, E. F. Castillo and O. Almazan, *Renewable Sustainable Energy Rev.*, 2009, 13, 1275–1287.

66. N. H. Tran, J. R. Bartlett, G. S. K. Kannangara, A. S. Milev, H. Volk and M. A. Wilson, *Fuel*, 2010, 89, 265–274.

67. G. Knothe, *Green Chem.*, 2011, 13, 3048–3065 RSC.

68. A. Karmakar, S. Karmakar and S. Mukherjee, *Bioresour. Technol.*, 2010, 101, 7201–7210.

69. *The Biodiesel Handbook*, AOCS Publishing, 2005 Search PubMed.

70. T. P. Durrett, C. Benning and J. Ohlrogge, *Plant J.*, 2008, 54, 593–607.

71. E. N. Frankel, *Lipid Oxidation*, The Oily Press, Bridgewater, England, 2nd edn, 2005.

72. C. A. W. Allen, K. C. Watts, R. G. Ackman and M. J. Pegg, *Fuel*, 1999, 78, 1319–1326.

73. G. Knothe and K. R. Steidley, *Energy Fuels*, 2005, 19, 1192–1200.

74. G. Knothe, S. C. Cermak and R. L. Evangelista, *Energy Fuels*, 2009, 23, 1743–1747.

75. R. Radakovits, R. E. Jinkerson, A. Darzins and M. C. Posewitz, *Eukaryotic Cell*, 2010, 9, 486–501.

76. Y. Ono and T. Baba, *Catal. Today*, 1997, 38, 321–337.

77. H. Hattori, *Chem. Rev.*, 1995, 95, 537–558.

78. M. C. G. Albuquerque, D. C. S. Azevedo, C. L. Cavalcante Jr, J. Santamaría-González, J. M. Mérida-Robles, R. Moreno-Tost, E. Rodríguez-Castellón, A. Jiménez-López and P. Maireles-Torres, *J. Mol. Catal. A: Chem.*, 2009, 300, 19–24.

79. G. R. Peterson and W. P. Scarrah, *J. Am. Oil Chem. Soc.*, 1984, 61, 1593–1597.

80. M. Verziu, S. M. Coman, R. Richards and V. I. Parvulescu, *Catal. Today*, 2011, 167, 64–70.

81. M. López Granados, D. Martin Alonso, A. C. Alba-Rubio, R. Mariscal, M. Ojeda and P. Brettes, *Energy Fuels*, 2009, 23, 2259–2263.

82. M. Di Serio, R. Tesser, L. Casale, A. Dapos;Angelo, M. Trifuoggi and E. Santacesaria, *Top. Catal.*, 2010, 53, 811–819.

83. M. Su, R. Yang and M. Li, *Fuel*, 2013, 103, 398–407.

84. C. S. MacLeod, A. P. Harvey, A. F. Lee and K. Wilson, *Chem. Eng. J.*, 2008, 135, 63–70.

85. J. Montero, K. Wilson and A. Lee, *Top. Catal.*, 2010, 53, 737–745.

86. R. S. Watkins, A. F. Lee and K. Wilson, *Green Chem.*, 2004, 6, 335–340 RSC.

87. D. M. Alonso, R. Mariscal, M. L. Granados and P. Maireles-Torres, *Catal. Today*, 2009, 143, 167–171.

88. M. Verziu, B. Cojocaru, J. Hu, R. Richards, C. Ciuculescu, P. Filip and V. I. Parvulescu, *Green Chem.*, 2008, 10, 373–381 RSC.

89. K. Zhu, J. Hu, C. Kübel and R. Richards, *Angew. Chem., Int. Ed.*, 2006, 45, 7277–7281.

90. J. M. Montero, P. Gai, K. Wilson and A. F. Lee, *Green Chem.*, 2009, 11, 265–268 RSC.

91. J. M. Montero, D. R. Brown, P. L. Gai, A. F. Lee and K. Wilson, *Chem. Eng. J.*, 2010, 161, 332–339.

92.  J. J. Woodford, C. M. A. Parlett, J.-P. Dacquin, G. Cibin, A. Dent, J. Montero, K. Wilson and A. F. Lee, *J. Chem. Technol. Biotechnol.*, 2014, 89, 73–80

93.  Y. L. Meng, B. Y. Wang, S. F. Li, S. J. Tian and M. H. Zhang, *Bioresour. Technol.*, 2013, 128, 305–309.

94.  W. Xie and L. Zhao, *Energy Convers. Manage.*, 2014, 79, 34–42.

95.  W. Xie and L. Zhao, *Energy Convers. Manage.*, 2013, 76, 55–62.

96.  E. Rashtizadeh, F. Farzaneh and Z. Talebpour, *Bioresour. Technol.*, 2013, 154C, 32–37.

97.  J.-X. Wang, K.-T. Chen, J.-S. Wu, P.-H. Wang, S.-T. Huang and C.-C. Chen, *Fuel Process. Technol.*, 2012, 104, 167–173.

98.  Y.-D. Long, F. Guo, Z. Fang, X.-F. Tian, L.-Q. Jiang and F. Zhang, *Bioresour. Technol.*, 2011, 102, 6884–6886.

99.  Y.-D. Long, Z. Fang, T.-C. Su and Q. Yang, *Appl. Energy*, 2014, 113, 1819–1825.

100. T. M. Barnard, N. E. Leadbeater, M. B. Boucher, L. M. Stencel and B. A. Wilhite, *Energy Fuels*, 2007, 21, 1777–1781.

101. X.-f. Li, Y. Zuo, Y. Zhang, Y. Fu and Q.-x. Guo, *Fuel*, 2013, 113, 435–442.

102. X. Liang, S. Gao, H. Wu and J. Yang, *Fuel Process. Technol.*, 2009, 90, 701–704.

103. H. Wu, J. Zhang, Y. Liu, J. Zheng and Q. Wei, *Fuel Process. Technol.*, 2014, 119, 114–120.

104. Y. Li, W. Zhang, L. Zhang, Q. Yang, Z. Wei, Z. Feng and C. Li, *J. Phys. Chem. B*, 2004, 108, 9739–9744.

105. W. Xie and M. Fan, *Chem. Eng. J.*, 2014, 239, 60–67.

106. K.-T. Chen, J.-X. Wang, Y.-M. Dai, P.-H. Wang, C.-Y. Liou, C.-W. Nien, J.-S. Wu and C.-C. Chen, *J. Taiwan Inst. Chem. Eng.*, 2013, 44, 622–629.

107. F. A. Dawodu, O. Ayodele, J. Xin, S. Zhang and D. Yan, *Appl. Energy*, 2014, 114, 819–826.

108. X. Fu, D. Li, J. Chen, Y. Zhang, W. Huang, Y. Zhu, J. Yang and C. Zhang, *Bioresour. Technol.*, 2013, 146, 767–770.

109. B. L. A. Prabhavathi Devi, T. Vijai Kumar Reddy, K. Vijaya Lakshmi and R. B. N. Prasad, *Bioresour. Technol.*, 2013, 153, 370–373 CrossRef PubMed.

110. Y. Shen, P. Zhao and Q. Shao, *Microporous Mesoporous Mater.*, 2014, 188, 46–76.

111. K. Gombotz, R. Parette, G. Austic, D. Kannan and J. V. Matson, *Fuel*, 2012, 92, 9–15.

112. A. Molaei Dehkordi and M. Ghasemi, *Fuel Process. Technol.*, 2012, 97, 45–51.

113. N. Santiago-Torres, I. C. Romero-Ibarra and H. Pfeiffer, *Fuel Process. Technol.*, 2014, 120, 34–39.

114. G. G. Santillán-Reyes and H. Pfeiffer, *Int. J. Greenhouse Gas Control*, 2011, 5, 1624–1629.

115. P. Hernández-Hipólito, M. García-Castillejos, E. Martínez-Klimova, N. Juárez-Flores, A. Gómez-Cortés and T. E. Klimova, *Catal. Today*, 2014, 220–222, 4–11.

116. D. G. Cantrell, L. J. Gillie, A. F. Lee and K. Wilson, *Appl. Catal., A*, 2005, 287, 183–190.

117. M. Di Serio, M. Ledda, M. Cozzolino, G. Minutillo, R. Tesser and E. Santacesaria, *Ind. Eng. Chem. Res.*, 2006, 45, 3009–3014

118. F. Cavani, F. Trifirò and A. Vaccari, *Catal. Today*, 1991, 11, 173–301.

119. N. Barakos, S. Pasias and N. Papayannakos, *Bioresour. Technol.*, 2008, 99, 5037–5042.

120. J. M. Fraile, N. García, J. A. Mayoral, E. Pires and L. Roldán, *Appl. Catal., A*, 2009, 364, 87–94.

121. H. E. Cross and D. R. Brown, *Catal. Commun.*, 2010, 12, 243–245.

122. T. Hibino and M. Kobayashi, *J. Mater. Chem.*, 2005, 15, 653–656 RSC.

123. J. M. Hidalgo, C. Jiménez-Sanchidrián and J. R. Ruiz, *Appl. Catal., A*, 2014, 470, 311–317.

124. J. J. Creasey, A. Chieregato, J. C. Manayil, C. M. A. Parlett, K. Wilson and A. F. Lee, *Catal. Sci. Technol.*, 2014, 4, 861–870.

125. E. Géraud, V. Prévot, J. Ghanbaja and F. Leroux, *Chem. Mater.*, 2005, 18, 238–240.

126. E. Géraud, S. Rafqah, M. Sarakha, C. Forano, V. Prevot and F. Leroux, *Chem. Mater.*, 2007, 20, 1116–1125.

127. J. J. Woodford, J.-P. Dacquin, K. Wilson and A. F. Lee, *Energy Environ. Sci.*, 2012, 5, 6145–6150.

128. M. N. V. Ravi Kumar, U. Bakowsky and C. M. Lehr, *Biomaterials*, 2004, 25, 1771–1777.

129. D. Highley, A. Bloodworth and R. Bate, *Dolomite-Mineral planning factsheet*, British Geological Survey, 2006.

130. D. Sutton, B. Kelleher and J. R. H. Ross, *Fuel Process. Technol.*, 2001, 73, 155–173.

131. K. Wilson, C. Hardacre, A. F. Lee, J. M. Montero and L. Shellard, *Green Chem.*, 2008, 10, 654–659 RSC.

132. O. Ilgen, *Fuel Process. Technol.*, 2011, 92, 452–455 CrossRef CAS PubMed.

133. Z. A. Shajaratun Nur, Y. H. Taufiq-Yap, M. F. Rabiah Nizah, S. H. Teo, O. N. Syazwani and A. Islam, *Energy Convers. Manage.*, 2014, 78, 738–744.

134. P. Zhang, Q. Han, M. Fan and P. Jiang, *Fuel*, 2014, 124, 66–72.

135. E. Karimi, I. F. Teixeira, L. P. Ribeiro, A. Gomez, R. M. Lago, G. Penner, S. W. Kycia and M. Schlaf, *Catal. Today*, 2012, 190, 73–88.

136. E. Karimi, A. Gomez, S. W. Kycia and M. Schlaf, *Energy Fuels*, 2010, 24, 2747–2757.

137. K. Narasimharao, D. R. Brown, A. F. Lee, A. D. Newman, P. F. Siril, S. J. Tavener and K. Wilson, *J. Catal.*, 2007, 248, 226–234.

138. K. Suwannakarn, E. Lotero, K. Ngaosuwan and J. G. Goodwin, *Ind. Eng. Chem. Res.*, 2009, 48, 2810–2818.

139. M. Kouzu, A. Nakagaito and J.-s. Hidaka, *Appl. Catal., A*, 2011, 405, 36–44.

140. D. Zhao, Q. Huo, J. Feng, B. F. Chmelka and G. D. Stucky, *J. Am. Chem. Soc.*, 1998, 120, 6024–6036.

141. I. K. Mbaraka and B. H. Shanks, *J. Catal.*, 2005, 229, 365–373.

142. J. A. Melero, L. F. Bautista, G. Morales, J. Iglesias and D. Briones, *Energy Fuels*, 2008, 23, 539–547.

143. X.-R. Chen, Y.-H. Ju and C.-Y. Mou, *J. Phys. Chem. C*, 2007, 111, 18731–18737.

144. I. K. Mbaraka, D. R. Radu, V. S. Y. Lin and B. H. Shanks, *J. Catal.*, 2003, 219, 329–336.

145. D. Chen, Z. Li, Y. Wan, X. Tu, Y. Shi, Z. Chen, W. Shen, C. Yu, B. Tu and D. Zhao, *J. Mater. Chem.*, 2006, 16, 1511–1519 RSC.

146. L. Cao, T. Man and M. Kruk, *Chem. Mater.*, 2009, 21, 1144–1153 CrossRef.

147. A. Martin, G. Morales, F. Martinez, R. van Grieken, L. Cao and M. Kruk, *J. Mater. Chem.*, 2010, 20, 8026–8035 RSC.

148. P. Zeigermann, S. Naumov, S. Mascotto, J. Kärger, B. M. Smarsly and R. Valiullin, *Langmuir*, 2012, 28, 3621–3632.

149. A. Vinu, N. Gokulakrishnan, V. V. Balasubramanian, S. Alam, M. P. Kapoor, K. Ariga and T. Mori, *Chem. – Eur. J.*, 2008, 14, 11529–11538.

150. C. Pirez, J.-M. Caderon, J.-P. Dacquin, A. F. Lee and K. Wilson, *ACS Catal.*, 2012, 2, 1607–1614.

151. D. E. López, J. G. Goodwin Jr, D. A. Bruce and E. Lotero, *Appl. Catal., A*, 2005, 295, 97–105.

152. W. W. Mar and E. Somsook, *J. Oleo Sci.*, 2013, 62, 435–442.

153. S.-Y. Chen, T. Mochizuki, Y. Abe, M. Toba and Y. Yoshimura, *Appl. Catal., B*, 2014, 148–149, 344–356.

154. J. A. Melero, J. Iglesias and G. Morales, *Green Chem.*, 2009, 11, 1285–1308 RSC.

155. A. Carrero, G. Vicente, R. Rodríguez, M. Linares and G. L. del Peso, *Catal. Today*, 2011, 167, 148–153.

156. J. Y. Ying, C. P. Mehnert and M. S. Wong, *Angew. Chem., Int. Ed.*, 1999, 38, 56–77.

157. Y. Lu, *Angew. Chem., Int. Ed.*, 2006, 45, 7664–7667 CrossRef CAS PubMed.

158. S. Garg, K. Soni, G. M. Kumaran, R. Bal, K. Gora-Marek, J. K. Gupta, L. D. Sharma and G. M. Dhar, *Catal. Today*, 2009, 141, 125–129

159. S. Gheorghiu and M.-O. Coppens, *AIChE J.*, 2004, 50, 812–820 CrossRef CAS Search PubMed.

160. X. Zhang, F. Zhang and K.-Y. Chan, *Mater. Lett.*, 2004, 58, 2872–2877.

161. J.-H. Sun, Z. Shan, T. Maschmeyer and M.-O. Coppens, *Langmuir*, 2003, 19, 8395–8402.

162. J.-P. Dacquin, J. r. m. Dhainaut, D. Duprez, S. b. Royer, A. F. Lee and K. Wilson, *J. Am. Chem. Soc.*, 2009, 131, 12896–12897.

163. J. Dhainaut, J.-P. Dacquin, A. F. Lee and K. Wilson, *Green Chem.*, 2010, 12, 296–303 RSC.

164. R. Hong, T. Pan, J. Qian and H. Li, *Chem. Eng. J.*, 2006, 119, 71–81.

165. M. L. Curri, R. Comparelli, P. D. Cozzoli, G. Mascolo and A. Agostiano, *Mater. Sci. Eng., C*, 2003, 23, 285–289.

166. G. P. Fotou and S. E. Pratsinis, *Chem. Eng. Commun.*, 1996, 151, 251–260.

167. S. Chakrabarti and B. Dutta, *J. Hazard. Mater.*, 2004, 112, 269–278.

168. H. Yoshida, S. Takashi, C. Murata and T. Hattori, *J. Catal.*, 2003, 220, 226–232.

169. G. Corro, U. Pal and N. Tellez, *Appl. Catal., B*, 2013, 129, 39–47

170. G. Corro, F. Bañuelos, E. Vidal and S. Cebada, *Fuel*, 2014, 115, 625–628.

171. C. Pirez, K. Wilson and A. F. Lee, *Green Chem.*, 2014, 16, 197–202 RSC.

172. N. Mizuno and M. Misono, *Chem. Rev.*, 1998, 98, 199–218.

173. I. V. Kozhevnikov, *Chem. Rev.*, 1998, 98, 171–198.

174. A. D. Newman, D. R. Brown, P. Siril, A. F. Lee and K. Wilson, *Phys. Chem. Chem. Phys.*, 2006, 8, 2893–2902 RSC.

175. A. D. Newman, A. F. Lee, K. Wilson and N. A. Young, *Catal. Lett.*, 2005, 102, 45–50.

176. L. Pesaresi, D. R. Brown, A. F. Lee, J. M. Montero, H. Williams and K. Wilson, *Appl. Catal., A*, 2009, 360, 50–58.

177. X. Duan, Y. Liu, Q. Zhao, X. Wang and S. Li, *RSC Adv.*, 2013, 3, 13748–13755 RSC.

178. S. Singh and A. Patel, *J. Cleaner Prod.*, 2014, 72, 46–56

179. P. Xia, F. Liu, C. Wang, S. Zuo and C. Qi, *Catal. Commun.*, 2012, 26, 140–143.

180. W. Liu, P. Yin, X. Liu, W. Chen, H. Chen, C. Liu, R. Qu and Q. Xu, *Energy Convers. Manage.*, 2013, 76, 1009–1014.

181. R. Tesser, M. Di Serio, L. Casale, L. Sannino, M. Ledda and E. Santacesaria, *Chem. Eng. J.*, 2010, 161, 212–222.

182. X. Liang, *Ind. Eng. Chem. Res.*, 2013, 52, 6894–6900

183. D. Zeng, S. Liu, W. Gong, G. Wang, J. Qiu and H. Chen, *Appl. Catal., A*, 2014, 469, 284–289.

184. C. A. Deshmane, M. W. Wright, A. Lachgar, M. Rohlfing, Z. Liu, J. Le and B. E. Hanson, *Bioresour. Technol.*, 2013, 147, 597–604.

185. D. D. Chabukswar, P. K. K. S. Heer and V. G. Gaikar, *Ind. Eng. Chem. Res.*, 2013, 52, 7316–7326.

186. C. Poonjarernsilp, N. Sano and H. Tamon, *Appl. Catal., B*, 2014, 147, 726–732.

187. F. H. Alhassan, R. Yunus, U. Rashid, K. Sirat, A. Islam, H. V. Lee and Y. H. Taufiq-Yap, *Appl. Catal., A*, 2013, 456, 182–187.

188. W. Xie and T. Wang, *Fuel Process. Technol.*, 2013, 109, 150–155.

189. R. Sheikh, M.-S. Choi, J.-S. Im and Y.-H. Park, *J. Ind. Eng. Chem.*, 2013, 19, 1413–1419.

190. M. Farooq, A. Ramli and D. Subbarao, *J. Cleaner Prod.*, 2013, 59, 131–140.

191. A. Talebian-Kiakalaieh, N. A. S. Amin and H. Mazaheri, *Appl. Energy*, 2013, 104, 683–710.

192. S. Yan, C. DiMaggio, S. Mohan, M. Kim, S. Salley and K. Y. S. Ng, *Top. Catal.*, 2010, 53, 721–736.

193. L. Peng, A. Philippaerts, X. Ke, J. Van Noyen, F. De Clippel, G. Van Tendeloo, P. A. Jacobs and B. F. Sels, *Catal. Today*, 2010, 150, 140–146.

194. P. S. Sreeprasanth, R. Srivastava, D. Srinivas and P. Ratnasamy, *Appl. Catal., A*, 2006, 314, 148–159.

195. Z. Helwani, M. R. Othman, N. Aziz, J. Kim and W. J. N. Fernando, *Appl. Catal., A*, 2009, 363, 1–10.

196. S. Miao and B. H. Shanks, *Appl. Catal., A*, 2009, 359, 113–120.

197. I. Jiménez-Morales, J. Santamaría-González, P. Maireles-Torres and A. Jiménez-López, *Appl. Catal., B*, 2011, 105, 199–205.

198. Q. Shu, J. Gao, Z. Nawaz, Y. Liao, D. Wang and J. Wang, *Appl. Energy*, 2010, 87, 2589–2596.

199. I. M. Atadashi, M. K. Aroua, A. R. Abdul Aziz and N. M. N. Sulaiman, *Renewable Sustainable Energy Rev.*, 2012, 16, 3456–3470.

200. D. Kusdiana and S. Saka, *Bioresour. Technol.*, 2004, 91, 289–295.

201. J. K. Satyarthi, D. Srinivas and P. Ratnasamy, *Energy Fuels*, 2010, 24, 2154–2161.

202. Y. Liu, E. Lotero and J. G. Goodwin Jr, *J. Catal.*, 2006, 243, 221–228.

203. D. M. Alonso, M. L. Granados, R. Mariscal and A. Douhal, *J. Catal.*, 2009, 262, 18–26.

204. K. Wilson, A. Rénson and J. H. Clark, *Catal. Lett.*, 1999, 61, 51–55.

205. B. Rác, P. Hegyes, P. Forgo and Á. Molnár, *Appl. Catal., A*, 2006, 299, 193–201.

206. Q. Yang, J. Liu, J. Yang, M. P. Kapoor, S. Inagaki and C. Li, *J. Catal.*, 2004, 228, 265–272.

207. Q. Yang, M. P. Kapoor, N. Shirokura, M. Ohashi, S. Inagaki, J. N. Kondo and K. Domen, *J. Mater. Chem.*, 2005, 15, 666–673 RSC.

208. G. Morales, G. Athens, B. F. Chmelka, R. van Grieken and J. A. Melero, *J. Catal.*, 2008, 254, 205–217.

209. R. Sánchez-Vázquez, C. Pirez, J. Iglesias, K. Wilson, A. F. Lee and J. A. Melero, *ChemCatChem*, 2013, 5, 994–1001.

210. D. Margolese, J. A. Melero, S. C. Christiansen, B. F. Chmelka and G. D. Stucky, *Chem. Mater.*, 2000, 12, 2448–2459.

211. I. Díaz, C. Márquez-Alvarez, F. Mohino, J. N. Pérez-Pariente and E. Sastre, *J. Catal.*, 2000, 193, 283–294.

212. J.-P. Dacquin, H. E. Cross, D. R. Brown, T. Duren, J. J. Williams, A. F. Lee and K. Wilson, *Green Chem.*, 2010, 12, 1383–1391 RSC.

213. C. Schumacher, J. Gonzalez, P. A. Wright and N. A. Seaton, *J. Phys. Chem. B*, 2005, 110, 319–333.

214. D. Zuo, J. Lane, D. Culy, M. Schultz, A. Pullar and M. Waxman, *Appl. Catal., B*, 2013, 129, 342–350.

215. L. Sherry and J. A. Sullivan, *Catal. Today*, 2011, 175, 471–476

216. J. A. Melero, R. van Grieken and G. Morales, *Chem. Rev.*, 2006, 106, 3790–3812.

217. A. P. S. Chouhan and A. K. Sarma, *Renewable Sustainable Energy Rev.*, 2011, 15, 4378–4399.

218. A. Macario, G. Giordano, B. Onida, D. Cocina, A. Tagarelli and A. M. Giuffrè, *Appl. Catal., A*, 2010, 378, 160–168.

219. D. Srinivas and J. Satyarthi, *Catal. Surv. Asia*, 2011, 15, 145–160

220. F. Yan, Z. Yuan, P. Lu, W. Luo, L. Yang and L. Deng, *Renewable Energy*, 2011, 36, 2026–2031.

221. T. Nakato, M. Kimura, S.-I. Nakata and T. Okuhara, *Langmuir*, 1998, 14, 319–325.

222. A. Drelinkiewicz, Z. Kalemba-Jaje, E. Lalik and R. Kosydar, *Fuel*, 2014, 116, 760–771.

223. G. Morales, R. van Grieken, A. Martín and F. Martínez, *Chem. Eng. J.*, 2010, 161, 388–396.

224. I. Noshadi, R. Kumar, B. Kanjilal, R. Parnas, H. Liu, J. Li and F. Liu, *Catal. Lett.*, 2013, 143, 792–797.

225. L. Geng, G. Yu, Y. Wang and Y. Zhu, *Appl. Catal., A*, 2012, 427–428, 137–144.

226. R. Liu, X. Wang, X. Zhao and P. Feng, *Carbon*, 2008, 46, 1664–1669.

227. B. Chang, J. Fu, Y. Tian and X. Dong, *J. Phys. Chem. C*, 2013, 117, 6252–6258.

228. M. Kotwal, A. Kumar and S. Darbha, *J. Mol. Catal. A: Chem.*, 2013, 377, 65–73.

229. L. Deng, T. Tan, F. Wang and X. Xu, *Eur. J. Lipid Sci. Technol.*, 2003, 105, 727–734.

230. M. Iso, B. Chen, M. Eguchi, T. Kudo and S. Shrestha, *J. Mol. Catal. B: Enzym.*, 2001, 16, 53–58.

231. C.-H. Liu, C.-C. Huang, Y.-W. Wang, D.-J. Lee and J.-S. Chang, *Appl. Energy*, 2012, 100, 41–46.

232. N. Dizge, B. Keskinler and A. Tanriseven, *Biochem. Eng. J.*, 2009, 44, 220–225.

233. L. Guerreiro, P. M. Pereira, I. M. Fonseca, R. M. Martin-Aranda, A. M. Ramos, J. M. L. Dias, R. Oliveira and J. Vital, *Catal. Today*, 2010, 156, 191–197.

234. X. Liu, H. He, Y. Wang, S. Zhu and X. Piao, *Fuel*, 2008, 87, 216–221.

235. S. Yan, H. Lu and B. Liang, *Energy Fuels*, 2007, 22, 646–651.

236. Q. Yang, J. Liu, L. Zhang and C. Li, *J. Mater. Chem.*, 2009, 19, 1945–1955 RSC.

237. B. Karimi, H. M. Mirzaei and A. Mobaraki, *Catal. Sci. Technol.*, 2012, 2, 828–834.

238. M. Bender, *Bioresour. Technol.*, 1999, 70, 81–87

239. T. Sakai, A. Kawashima and T. Koshikawa, *Bioresour. Technol.*, 2009, 100, 3268–3276.

240. E. F. Aransiola, T. V. Ojumu, O. O. Oyekola, T. F. Madzimbamuto and D. I. O. Ikhu-Omoregbe, *Biomass Bioenergy*, 2014, 61, 276–297.

241. M. B. Tasić, O. S. Stamenković and V. B. Veljković, *Energy Convers. Manage.*, 2014, 84, 405–413.

242. Y. Cheng, Y. Feng, Y. Ren, X. Liu, A. Gao, B. He, F. Yan and J. Li, *Bioresour. Technol.*, 2012, 113, 65–72.

243. A. A. Kulkarni, K.-P. Zeyer, T. Jacobs and A. Kienle, *Ind. Eng. Chem. Res.*, 2007, 46, 5271–5277.

244. Ó. de la Iglesia, R. Mallada, M. Menéndez and J. Coronas, *Chem. Eng. J.*, 2007, 131, 35–39.

245. A. A. Kiss, A. C. Dimian and G. Rothenberg, *Energy Fuels*, 2007, 22, 598–604.

246. C. Buchaly, P. Kreis and A. Górak, *Ind. Eng. Chem. Res.*, 2011, 51, 891–899.

247. Z. Qiu, L. Zhao and L. Weatherley, *Chemical Engineering and Processing: Process Intensification*, 2010, 49, 323–330.

248. G. L. Maddikeri, A. B. Pandit and P. R. Gogate, *Ind. Eng. Chem. Res.*, 2012, 51, 14610–14628.

249. X. Ni, M. R. Mackley, A. P. Harvey, P. Stonestreet, M. H. I. Baird and N. V. Rama Rao, *Chem. Eng. Res. Des.*, 2003, 81, 373–383

250. A. N. Phan, A. P. Harvey and V. Eze, *Chem. Eng. Technol.*, 2012, 35, 1214–1220.

251. R. G. Nelson and S. A. Hower, Sixth national bioenergy conference, 1994.

252. S. Baroutian, M. K. Aroua, A. A. Raman and N. M. Sulaiman, *Bioresour. Technol.*, 2011, 102, 1095–1102.

253. H. Falahati and A. Y. Tremblay, *Fuel*, 2012, 91, 126–133.

254. W. Xu, L. Gao, S. Wang and G. Xiao, *Bioresour. Technol.*, 2014, 159, 286–291.

255. P. Cao, A. Y. Tremblay, M. A. Dubé and K. Morse, *Ind. Eng. Chem. Res.*, 2007, 46, 52–58.

256. P. Cao, A. Y. Tremblay and M. A. Dubé, *Ind. Eng. Chem. Res.*, 2009, 48, 2533–2541.

257. P. Lozano, J. M. Bernal and M. Vaultier, *Fuel*, 2011, 90, 3461–3467.

258. X. Wang, X. Liu, C. Zhao, Y. Ding and P. Xu, *Bioresour. Technol.*, 2011, 102, 6352–6355.

259. D. Lv, W. Du, G. Zhang and D. Liu, *Process Biochem.*, 2010, 45, 446–450.

260. A. Bajaj, P. Lohan, P. N. Jha and R. Mehrotra, *J. Mol. Catal. B: Enzym.*, 2010, 62, 9–14.

261. L. A. Nelson, T. A. Foglia and W. N. Marmer, *J. Am. Oil Chem. Soc.*, 1996, 73, 1191–1195.

262. Y. Watanabe, Y. Shimada, A. Sugihara, H. Noda, H. Fukuda and Y. Tominaga, *J. Am. Oil Chem. Soc.*, 2000, 77, 355–360.

263. K. Bélafi-Bakó, F. Kovács, L. Gubicza and J. Hancsók, *Biocatal. Biotransform.*, 2002, 20, 437–439.

264. M. Lee, D. Lee, J. Cho, S. Kim and C. Park, *Appl. Biochem. Biotechnol.*, 2013, 171, 1118–1127.

265. K. T. Tan and K. T. Lee, *Renewable Sustainable Energy Rev.*, 2011, 15, 2452–2456.

266. V. G. Gude and G. E. Grant, *Appl. Energy*, 2013, 109, 135–144.

267. V. L. Gole and P. R. Gogate, *Chemical Engineering and Processing: Process Intensification*, 2012, 53, 1–9.

268. A. Mazubert, C. Taylor, J. Aubin and M. Poux, *Bioresour. Technol.*, 2014, 161, 270–279.

269. W. A. Wali, A. I. Al-Shamma'a, K. H. Hassan and J. D. Cullen, *J. Process Control*, 2012, 22, 1256–1272.

270. P. Chand, V. R. Chintareddy, J. G. Verkade and D. Grewell, *Energy Fuels*, 2010, 24, 2010–2015.

271. B. Salamatinia, H. Mootabadi, S. Bhatia and A. Z. Abdullah, *Fuel Process. Technol.*, 2010, 91, 441–448.

272. H. A. Choudhury, S. Chakma and V. S. Moholkar, *Ultrason. Sonochem.*, 2014, 21, 169–181.

273. P. Khemthong, C. Luadthong, W. Nualpaeng, P. Changsuwan, P. Tongprem, N. Viriya-empikul and K. Faungnawakij, *Catal. Today*, 2012, 190, 112–116.

Chapter

# 2

# Contrast and Synergy between Electrocatalysis and Heterogeneous Catalysis

Andrzej Wieckowski[1] and Matthew Neurock[2]

[1]Department of Chemistry, University of Illinois at Urbana-Champaign, Urbana, IL 61801, USA

[2]Departments of Chemical Engineering and Chemistry, University of Virginia, Charlottesville, VA 22904-4741, USA

## ABSTRACT

The advances in spectroscopy and theory that have occurred over the past two decades begin to provide detailed in situ resolution of the molecular transformations that occur at both gas/metal as well as aqueous/metal interfaces. These advances begin to allow for a more direct comparison of heterogeneous catalysis and electrocatalysis. Such comparisons become important, as many of the current energy conversion strategies involve catalytic and electrocatalytic processes that occur at fluid/solid interfaces and display very similar characteristics. Herein, we compare and contrast a few different catalytic and electrocatalytic systems to elucidate the principles that cross-cut both areas and establish characteristic differences between the two with the hope of advancing both areas.

# INTRODUCTION

Electrocatalysis and heterogeneous catalysis are closely related in that they involve well-controlled sequences of elementary bond-breaking and making processes and share many common mechanistic principles in the transformation of molecules over supported metal and metal oxide catalysts. While there are many areas of synergy between the two, including the materials that are used and the available reaction pathways and mechanisms, there are also well-established differences [1–4]. Heterogeneous catalysis has often celebrated more detailed insights into reaction mechanisms than electrocatalysis due to the advances in spectroscopy and theory of the gas/solid interface as compared to the more complex aqueous/solid interface in electrocatalysis. As such electrocatalysis has often followed from the mechanistic advances derived from gas phase heterogeneous catalysis. Many of the current efforts in heterogeneous catalysis, however, are focused on energy conversion strategies involving catalytic transformations which proceed at the fluid/solid interface and, as a consequence, are now closely following the leads from electrocatalysis. A number of common mechanistic principles and features are beginning to emerge between the two fields. Understanding the synergies as well as the differences between catalysis and electrocatalysis should thus enable advances in the science and application for both areas. Herein, we compare and contrast some of the fundamental mechanistic constructs as well as the practical applications for electrocatalysis and heterogeneous catalysis. More specifically, we focus on metal catalyzed oxidation processes.

# GENERAL COMPARISONS BETWEEN CATALYSIS AND ELECTROCATALY-SIS

At the macroscopic level, many of the catalytic materials that are used in catalysis and electrocatalysis are very similar in that they involve supported metal particles, where the interaction between the metal and support is critical to catalyst performance as well as catalyst stability. The metal or metal oxide/support interface can result in sites with unique structural or electronic characteristics, novel bifunctional sites, and/or sites that promote proton and electron transfer. The nature and the strength of the bonds between the metal and the support control the stability of these materials and their resilience to harsh reaction environments.

The characterization of the electronic and atomic structure of the metal and the support in both catalysis as well as electrocatalysis is typically carried out through the use of extended X-ray absorption spectroscopy (XAFS), electron microscopy, X-ray (XPS) and ultraviolet photoelectron spectroscopy (UPS). In addition, many of the most active metals used in electrocatalysis are very often the same as those used in heterogeneous catalysis. For example, Pt and other group VIII metals are known to be very active in the electrocatalytic oxidation of alcohols and the reduction of oxygen in fuel cells, automotive exhaust catalysis and hydrogenolysis, and hydrogenation catalysis in the conversion of petroleum and renewable resources. This is predominantly the result of the well-established Sabatier's Principle which suggests that the metals in middle of the periodic table demonstrate an optimal metal-adsorbate bond strengths necessary to balance surface reaction steps and product desorption steps [5–9].

In addition to similarities, there are well-established differences between traditional gas phase heterogeneous catalysis and electrocatalysis. Perhaps the greatest difference between the two relates to the unique reaction environments in which they are carried out. The gas phase catalytic environment is far less complex than that of the electrified water/metal interface for electrocatalytic systems, thus allowing for more detailed spectroscopic characterization of the working surface intermediates, application of ultrahigh vacuum experiments as well as direct comparisons with theoretical simulations on model surfaces. The presence of solution, ions, charged interfaces, complex surface potentials, and electric fields present in electrocatalytic systems can all act to significantly change the surface chemistry and catalysis that occurs in these environments. These interfaces tend to significantly promote polar reactions and direct heterolytic bond activation steps which would otherwise be unstable and not occur in gas phase catalytic systems. The electrochemical environment, however, is typically much harsher and deleterious to catalyst stability than that found in gas phase catalysis. The dissolution of the metal and the support are thus important concerns for electrocatalytic processes as these steps are enhanced under electrochemical conditions. In addition, the presence of electrolyte often enhances or impedes catalytic kinetics and within certain potential regions can result in poisoning of the surface.

While there are important differences between electrocatalysis and catalysis that result from the presence of solution, counter ions, and electric fields, Nørskov [8–12], Anderson [13–18], and others [1, 6] have been able to model the electrochemical systems by simply carrying out gas phase calculations on well-defined model clusters and surfaces and adding in the critical features that influence the surface chemistry such as local water molecules as well as the influence of potential. This is an important step in that one has the ability to not only understand but begin to tune the reaction chemistry. Understanding the similarities and differences of the molecular transformations that occur at ultrahigh vacuum conditions and electrochemical conditions will undoubtedly drive advances in the development of catalytic and electrocatalytic materials and processes.

In addition to the scientific issues, there are also a number of important technological differences in the "infrastructure" that supports the heterogeneous and electrocatalysis communities. Most of electrocatalysis appears to be centered around fuel cells and more specifically proton exchange membrane (PEM) fuel cells carry out the oxidation of hydrogen, oxygenates, or hydrocarbon molecules to $CO_2$ and the reduction oxygen to water [19]. This is in clear contrast to heterogeneous catalysis, which spans a wide and diverse range of different molecules with very rich chemistry and stands behind extensive chemical, automotive, petroleum, and pharmaceutical industries. In addition, there have been significant research and development investments in heterogeneous catalysis from the government as well as industry. Methane reforming, methane combustion, ammonia synthesis $NO_x$ conversion, and Fischer Tropsch synthesis, for example, are very large-scale processes which have no analogue in electrocatalysis. Furthermore, the future directions for PEM fuel cell catalysis is very specific with a strong focused effort on resolving the issues related to durability and maintaining catalytic activity for many years. This is quite different than the shorter lifetimes involved in most heterogeneous catalytic processes with the exception of automotive emission catalysis. In electrocatalysis, catalyst must be able to withstand the harsh operating conditions and operate effectively over the lifetime of the vehicle. Catalyst loss and deactivation tend to be quite severe in electrocatalysis due to the presence of solution, ions, and electric fields which not only lead to catalyst poisoning but also catalyst dissolution. This significantly limits the choice of catalytic materials to specific supported metals/alloys, metal oxides, and other stable inorganic materials such as chalcogenides or carbides. This is a very narrow range of possibilities as compared to what is typically practiced in the gas phase heterogeneous catalysis community. The long term durability, aggressive solution conditions (both high and low pH), as well as the cost tend to prevent other avenues available to gas phase catalysis to be applied in the field of electrocatalysis.

Despite the complexity and the challenges of the electrochemical environment and the differences outlined above, many of the fundamental constructs that govern gas phase catalysis are also integral to electrocatalysis. There have been a number of pioneering developments in spectroscopy, kinetic analyses, theory, and synthesis that have occurred over the past few decades that have clearly shown how traditional concepts from heterogeneous catalysis apply directly to electrocatalysis. This includes the elucidation of nature of the active site, competitive adsorption phenomena, the influence of alloys, promoters and poisons, structure sensitivity, surface oxidation state, particle size effects, and metal support interactions. A schematic representation of the complex metal solution interface that would exist in either the catalytic or electrocatalytic oxidation of glucose over a carbon-supported Pt Ru alloy cluster is shown in Figure 1. In a review that is now over ten years old, Jarvi and Stuve [20] elegantly described the direct link between some of the fundamental principles that control heterogeneous catalysis and electrocatalysis. This included the specific accounting of active sites, the identification of reaction intermediates in elucidating reaction mechanisms, and the role of poisons and modifiers and their influence on catalytic kinetics and aging. The authors nicely showed that the kinetics for catalysis and electrocatalysis were essentially the same and shared a common framework.

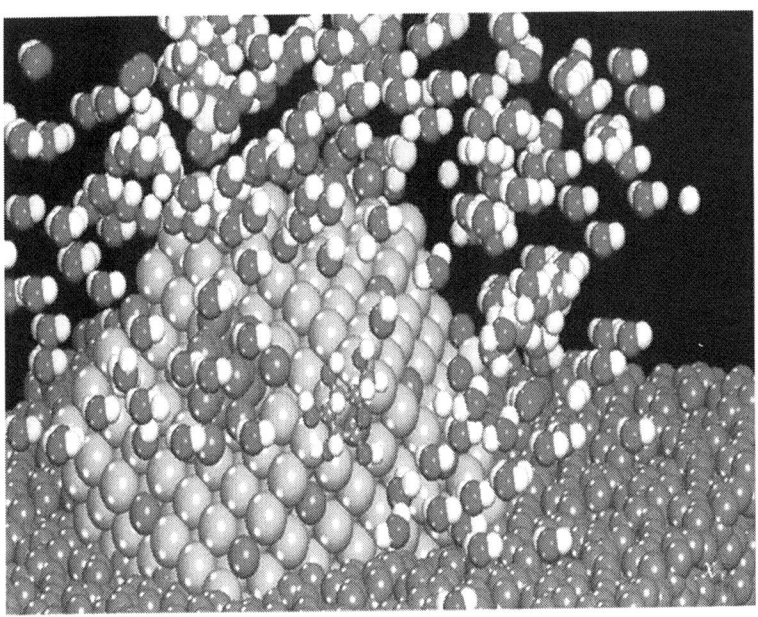

**Figure 1:** Schematic representation of the complex aqueous/metal interface involved in the catalytic and electrocatalytic oxidation of glucose over a PtRu alloy particle supported on carbon in the presence of electrolyte.

The complexity of the aqueous/metal interface and the inability to spectroscopically resolve molecular intermediates at this interface limited early electrocatalytic studies to measurements of macroscopic kinetics with little understanding of the elementary molecular transformations, the active sites or the influence the atomic and the electronic structure. This put electrocatalysis at a distinct disadvantage over traditional gas phase heterogeneous catalysis. The tremendous breakthroughs that have occurred over the past two decades in the ability to characterize the atomic structure of the working surface and reaction intermediates on the surface within the electrochemical environments have made it possible to begin to discuss elementary mechanisms and the influence of specific structural and composition parameters. Breakthroughs in broad-band spectroscopy [21], in situ electrochemical NMR [22–24], EXAFS [25–29], and surface-enhanced infrared spectroscopy [30, 31] are helping to significantly advance our understanding of electrocatalysis and in some cases begin to rival that in gas phase catalysis. These methods now allow for the direct insights into the nature of the active sites as well as surface intermediates under actual working conditions. Such insight has led to an exponential growth in the literature in the identification of active sites, reaction mechanisms, and rigorous structure-property relationships. Despite these important advances, the molecular resolution of intermediates under working conditions still presents a significant challenge and as such has been limited to only a few different intermediates and systems. In addition, these methods provide information on model systems that lack the complexity of the actual catalytic environment and often provide only a limited understanding.

In addition to these advances in spectroscopy, the past two decades have witnessed exponential increases in computational power and tremendous advances in theory and simulation methods. The development of density functional theory along with higher level ab initio wave function methods, and novel embedding methods, for example, has revolutionized our understanding adsorbate bonding and reactivity on well-defined surfaces and organometallic clusters. Due to limited computational resources, methods, and knowledge, most of these initial theoretical studies were focused on modeling ideal single-crystal surfaces under vacuum conditions. The insights and confidence gained from these initial efforts together with further increases in computational methods and nurturing from the experimental electrocatalysis community have helped to "seed" the exponential growth that has occurred in the development and the application of theory in electrocatalysis over the past ten years. Many of the initial developments were based on important insights into the electronic factors that controlled electrocatalytic reactivity. Andserson pioneered the development of reaction center model [13, 15, 18, 32–35], whereas Nørkov and colleagues [11, 12] developed a simple but elegant method that directly relates gas phase surface reaction energies to reaction energies at applied potentials. Schmickler et al. [36–38] developed a model Hamiltonian that appropriately captures bond-breaking and bond-making processes that occur over metal surfaces in electrochemical systems by combining fundamental electron transfer and solvent reorganization principles derived from Marcus theory, Newns Anderson theory on surface reactivity, and a tight binding theory.

These initial efforts were subsequently followed by ab initio-based simulation methods to follow chemistry within the aqueous metal interface and the direct relationship to electrocatalysis at applied potentials. There are now a number of rather sophisticated models that include the presence of solution, electrochemical potentials, applied fields, and actual electrolyte in modeling the electrocatalysis. Filhol, Taylor, and Neurock used explicit electrolyte or charge to establish the double layer at the surface [39–43]. The charged surfaces were then referenced to vacuum in order to establish the working potential. Otani et al. used DFT to describe the water metal interface and coupled this with an effective screening medium to represent to polarizable continuum [44, 45]. Jionnouchi and Anderson developed a similar approach by combined density functional theory and modified Poisson-Boltzmann theory [46]. Rossmeisl et al. [47] used explicit protons at the water/metal interface to establish the double-layer interface. While the models by Neurock, Otani, Anderson, and Nørskov differ in how they treat the double layer, they are providing more rigorous solutions to the electrochemical transformations that occur on electrode surfaces. It is important, however, to note that these approaches are at best semiquantitative due to limitations of fundamental accuracy of the quantum mechanical methods, modeling electron transfer reactions and simulating long time dynamics, the full reaction environment, or the millions of configurations needed for accurate statistical treatments. Vapor phase density functional calculations of bond energies and activation barriers are typically within the range of 0.1–0.2 eV accuracies but can have outliers [2, 48]. The simulation of electrochemical systems would at best only be 0.3 V. Despite these issues, theory has plaid and will likely continue to play a very valuable role in understanding and establishing trends.

The discussions that follow will look to theory only to provide insights rather than quantitative predictions. All of the simulations reported were carried out gradient corrected periodic DFT calculations using the Vienna Ab Initio Simulation Program (VASP) [49, 50] with four-layer metal slabs in the presence of solution, where the bottom two layers in the metal were held fixed to the bulk lattice positions of the metal. The coordinates of the metal atoms in the top two layers along with all of the atoms in the adsorbates as well as in the solution layer were fully optimized. Transition states were isolated using the nudged elastic band method with climbing [51, 52] followed by the dimer method [53]. The specific details are reported in the previous papers [39–43]. Rather than continue to discuss the obvious connections between electrocatalysis and gas phase heterogeneous catalysis, it is perhaps more interesting to discuss the growing efforts for carrying out heterogeneous catalysis in solution and new and emerging results that connect heterogeneous catalysis to well-established principles and phenomena in electrocatalysis. There has been an exponential growth over the past few years in carrying out heterogeneous catalytic reactions in solvents or aqueous media. This has been the result of the significant efforts to convert biomass into chemicals and fuels [54–59]. The carbohydrates that result from the breakdown of biomass are soluble in aqueous media and in addition can be catalytically converted at lower temperatures and much milder conditions than traditional gas phase processes. Similarly, many of the processes used in the selective hydrogenation and selective oxidation of fine chemical and pharmaceutical intermediates are also carried out in aqueous or solvent media that operate at lower temperatures to control both chemical as well as enantiomeric selectivities. Many of the catalyst performance and durability issues found in these systems have strong parallels to those found in electrocatalysis. Despite the similarities, there have been very few attempts to connect or compare the two.

# SPECIFIC COMPARISONS BETWEEN CATALYSIS AND ELECTROCATALYSIS: EXAMPLE SYSTEMS

## Metal Support Interface

Before discussing specific chemistry, we will first focus the metal-support interface in aqueous phase heterogeneous catalysis and electrocatalysis. While the lower temperatures used in aqueous phase catalysis help to control the reaction selectivity, the presence of water often leads to the hydrolysis of metal support bonds which can significantly limit the supports that can be used due to issues related to metal sintering and dissolution. Much of the initial work in the area of conversion of biorenewables was carried out over traditional transition metal catalysts (Pd, Ru, Pt, and Ni) and their supports including $SiO_2$, $Al_2O_3$, $TiO_2$, and high surface area carbons in order to identify active and selective materials [54–57,60]. There was little emphasis on the fundamental surface chemistry that occurred in the solution phase or the stability of these materials. Maris et al. [61, 62] were some of the first to identify the potential issues related to metal-support interactions under aqueous phase catalytic conditions. They showed that while Ru supported on $SiO_2$ leads to 100% selective hydrogenation of glucose to the sugar alcohol, the catalyst used was inherently unstable in aqueous media and resulted in the significant metal particle growth. Subsequent studies by Ketchie et al. [63] used in situ X-ray absorption spectroscopy to monitor the oxidation state of the Ru and follow the metal stability over a range of traditional supports including $SiO_2$, -$Al_2O_3$, carbon, and $TiO_2$. Significant sintering of Ru on both $SiO_2$ and $Al_2O_3$ supports occurred at the mild conditions associated with the conversion of biorenewables. Both the high surface area activated carbon and titania, on the other hand, were found to be stable supports for aqueous phase catalysis over a range of operating conditions. They showed that different carbons may behave differently and that great care must be taken to elucidate the nature of the metal-support interactions under actual process conditions. Most of the recent studies on the conversion of renewables as well as the hydrogenation of pharmaceutical intermediates are now carried out on activated high surface area carbon due to the stability issues.

Interestingly, many of these issues were resolved in the electrocatalysis community many years earlier as metal dissolution is one of the key issues that limit fuel cell durability. Carbon has been the preferred support throughout electrocatalysis as a result of the stability, durability, conductivity, and reactivity of the metal/carbon interface. More recent efforts have demonstrated that the introduction of titania can help to stabilize the metal/support interactions at the cathode for oxygen reduction. Despite the advances in both catalysis and electrocatalysis towards stabilizing the metal/support interface, it is clear that this is an important area which will require mechanistic insights into the fundamental processes that lead to dissolution and loss of metal and practical advances to solve the issues of durability.

# Oxidation of CO

## *Catalytic Oxidation of CO*

A second example in which heterogeneous catalysis and electrocatalysis are related is the recent discoveries concerning the unique catalytic activity of supported metal particles in the presence of an aqueous medium. Perhaps most evident is the work that has been carried out over supported Au. Up until 1987, gold was considered to be inert and inactive for catalysis. In a pioneering discovery, Haruta et al. [64, 65] demonstrated that nanometer-sized Au particles supported on $TiO_2$ were highly active for low-temperature CO oxidation in the gas phase. This work led to a tremendous number of follow-up studies aimed at understanding the mechanism by which this reaction proceeds and demonstrates the unique behavior of nanometer and subnanometer-sized Au particles in catalyzing a range of different reactions over $TiO_2$ as well as other supports [66, 67]. A number of possible explanations for the unique reactivity of Au have been presented in the literature [67] including quantum-size effects [68], increased coordinatively unsaturated edge and corner sites [69], the presence of cationic or anionic Au centers [70, 71], and unique sites at the $Au/TiO_2$ interface [72]. While the mechanism is still openly debated, much of the literature suggests that sites along the interface are responsible for the high catalytic activity. Some have speculated that the Au-Ti site pairs that result at the $Au/TiO_2$ interface stabilize the adsorption and activation of $O_2$ or the formation of hydroxyl intermediates both of which can catalyze these reactions [67]. The later idea is supported by the fact that the introduction of small amounts of water or base significantly promotes CO oxidation over $Au/TiO_2$ [73, 74]. Catalysis on these active 2–4 nm-sized Au clusters is thought to be quite different than that found in the electrocatalysis, as the former is carried out in the gas phase at low temperatures over small Au clusters without an applied potential or the presence of promoters. In addition, $O_2$ is the active oxidant in gas phase heterogeneous catalysis, whereas activated water or hydroxyl intermediates are thought to oxidize CO in electrocatalysis.

Kim et al. [75] and Sanchez-Castillo et al. [76] and Ketchie et al. [77, 78] later demonstrated that bulk Au as well as Au nanotubes were also very active in catalyzing CO oxidation if the reactions were carried out in an aqueous media. The rates over bulk Au were found to be over an order of magnitude higher in water than those in gas phase. While CO oxidation can proceed readily in the gas phase for nanometer clusters of Au on $TiO_2$, water is necessary for carrying the reactions out over bulk Au as well as Au supported on carbon.

These same aqueous phase CO oxidation rates were increased by up to 50 times upon increasing the pH of solution from neutral conditions (pH = 7) to basic conditions (pH = 14) and by over an order of magnitude from those in neutral solutions when the reactions were carried out in the presence of 0.5% $H_2O_2$ [76].

Dumesic et al. suggested that mechanism for CO oxidation in water may proceed via the formation of surface hydroxyl intermediates that can catalyze CO oxidation similar to the classic bifunctional mechanism suggested in electrocatalysis [79–82] and shown previously from theoretical calculations by Desai and Neurock [83, 84]. In this mechanism, water is activated on Ru sites within a PtRu surface alloy, whereas CO binds/blocks the Pt sites as is shown in Figure 2. CO oxidation proceeds via a nucleophilic attack of the OH (bound to Ru) on a neighboring CO (bound to Pt) coupled with the heterolytic splitting of O–H to form $CO_2$ as well as an electron and a proton. This mechanism would also help to explain the significant promotional effects that occur when the reaction is carried out in the presence of base both catalytically [76, 78] as well as electrocatalytically [85, 86].

**Figure 2**: DFT-calculated transition state for the electrocatalytic oxidation of CO on Pt (green spheres) by hydroxyls formed via the dissociation of water over Ru (yellow spheres). The resulting electron is transferred to the metal whereas the proton shuttles away from the surface via the water network (copyright Science [83]).

## *Electrocatalytic Oxidation of CO over Au*

The anodic oxidation of CO over Au/C was actually established in 1965, over two decades before the pioneering work of Haruta [87, 88]. The reaction readily proceeds over single crystal as well as polycrystalline Au electrodes at low temperature in alkaline media at potentials which are 0.5 V lower than those found with Pt which is used in most fuel cells. While CO oxidation readily occurs over single-crystal electrodes in acidic media, the rate is significantly enhanced when carried out in alkaline media. CO oxidation proceeds over Au(111) and Au(110) at potentials of 0.1 and 0.2 V RHE, respectively, which is 0.5 V lower than those recorded in acid media [88]. Similar differences for CO oxidation in acidic and alkaline media also exist for polycrystalline

Au. The reaction is thought to proceed via the coupling of CO* and OH* in a mechanism that is similar to that presented above in Figure 2. While the reaction proceeds in acidic media, the activation of water to form OH* on Au at lower potentials is difficult. The reaction is significantly faster in alkaline media as a result of increased formation of OH* at lower potentials and the stronger adsorption of CO at these lower potentials. The higher coverages of CO* and OH* in alkaline media further enhance the rate of reaction at the lower potentials. This shift in potential results in an increase in electron density at the metal surface which further enhances backdonation. Rodriguez et al. [88, 89] suggested that the increased CO adsorption further enhances the adsorption of OH and self-promotes the reaction.

Nearly all of the early electrocatalytic studies were carried out over bulk polycrystalline Au electrodes. The exceptional findings by Haruta et al. [64, 65] and the suggestions of the unique interface for nanoparticles of Au supported on $TiO_2$ prompted Hayden et al. [85, 86] to examine the electrocatalytic oxidation of CO over nanoparticles of Au on $TiO_2$. They demonstrated considerable enhancements in the electrocatalytic rates of CO oxidation for Au nanoparticles supported on $TiO_2$ even when the reaction is carried out in acidic media. The enhancement was attributed to the substrate-induced reactivity of Au as discussed below [85, 86].

## Particle Size Effects on CO Oxidation

The results from gas phase catalysis carried out over supported Au clusters indicate that the reaction is structure sensitive where the highest catalytic activity occurs for 2–4 nm sized particles supported on $TiO_2$ [66, 67]. The results in aqueous media, however, are not as clear. Ketchie et al. show a significant increase in the activity in moving from 42 nm down to 5 nm [77]. The activity on 2-3 nm-sized clusters, however, was more difficult to quantify as a result of difficulties in the synthesis of monodisperse particles [64, 65, 67]. Under electrochemical conditions, Hayden et al. [85, 86] showed that while polycrystalline Au electrodes readily oxidize CO to $CO_2$, they require overpotentials greater than 0.7 V as AuO forms and inhibits the reaction. They demonstrated significant structure sensitivity in the reaction, whereby 3 nm Au particles were much more active even at potentials as low as 0.3 V. Au particles smaller than 2.5 nm showed a significant decrease in activity as did particles larger than 3.5 nm. They speculate that the loss in activity below 3 nm is the result of quantum size effects rather than from irreversibly adsorbed oxygen. There appears to be clear similarities for particle size effects in comparing the results for CO oxidation under electrocatalytic conditions with those found for Au/TiO$_2$ in the gas phase heterogeneous catalysis. The reactions over Au/C in solution, however, were less conclusive.

## CO Oxidation Mechanisms

As was suggested earlier, the heterogeneous catalytic and the electrocatalytic oxidation of CO may proceed by common mechanisms or at least common features in the mechanism. We compare the mechanistic ideas for the catalytic and electrocatalytic oxidation of CO over both TiO$_2$ and carbon supports.

## (1) Catalytic and Electrocatalytic Oxidation of CO over TiO$_2$

The unique reactivity of nanometer-sized Au particles on TiO$_2$ used in both catalysis as well as electrocatalysis is thought to be dictated by sites at the Au/TiO$_2$ interface. We have recently shown that the Ti cations in direct proximity to the adsorbed Au become positively charged as a result of local charge transfer from Au to local Ti$^{5+}$ cations as is shown in Figure 3 [90]. This increase in charge along with the direct involvement of both Au and Ti$^{5+}$ atoms stabilize the bidentate adsorption of O$_2$. This increases the O$_2$ adsorption strength by nearly 60 kJ/mol and promotes its activation at these dual interface sites. This subsequently catalyzes the reaction between O$_2$ coadsorbed CO. While O$_2$ is not present in the electrocatalytic oxidation of CO, water can adsorb and activate at these same perimeter sites. DFT calculations indicate that the adsorption of water at this same charged interfacial Ti$^{5+}$ site is −104 kJ/mol, whereas water on a Ti$^{5+}$ site removed from Au and on a Au site near the support are significantly weaker at −77 kJ/mol and −8 kJ/mol, respectively. These Au-Ti perimeter sites likely attenuate the activation of water. The resulting intermediate (either activated water or surface hydroxyl groups) can then react with coadsorbed CO to form CO$_2$. This is consistent with the results from Hayden et al.

[85, 86] who showed a direct relationship between electrocatalytic activity and the perimeter of the particle and was able to rule out the influence of low coordinate Au atoms and quantum size effects. The differences between the gas phase reactions and the electrocatalytic reactions may simply be the nature of the active oxidant that forms. In the presence of water, the oxidant involved in catalysis and electrocatalysis is likely the same.

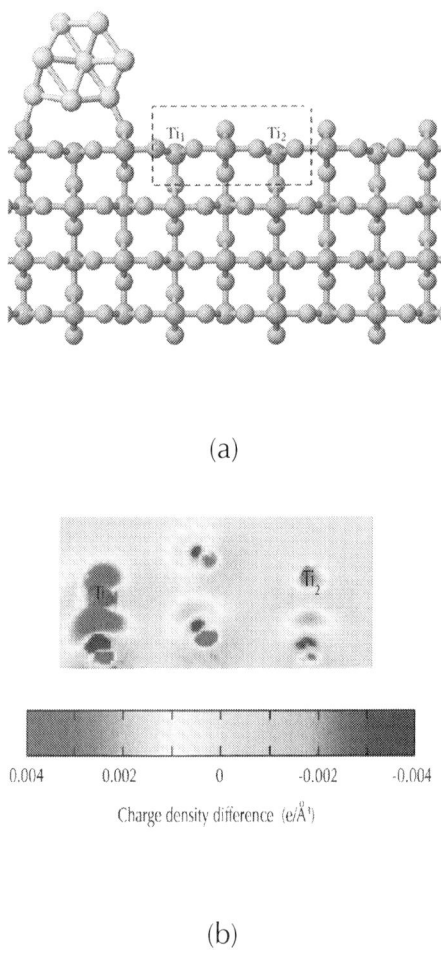

(a)

(b)

**Figure 3:** Charge density difference that results at $Ti^{5+}$ centers of $TiO_2$ adjacent to Au upon the adsorption of Au nanorods or clusters. (Copyright Science [90]).

## (2) Au/C the Effects of Water and Base

The heterogeneous oxidation of CO over bulk Au/C in water appears to proceed via a mechanism that involves the nucleophilic addition of OH groups present in solution or on the surface formed under reaction conditions to adsorbed CO*. This results in the formation of $CO_2$ and either adsorbed hydrogen or a proton and an electron.

$$CO^*+OH^* \rightarrow COOH^* \rightarrow CO_2(aq) +H^*+^* \qquad (1)$$

$$CO^*+OH^* \rightarrow COOH^* \rightarrow CO_2 (aq)+H^++e^-+^* \qquad (2)$$

The latter is identical to the mechanism typically proposed in electrocatalysis. The results for the heterogeneous CO oxidation in water over Au/C indicate that the reaction can proceed solely by the presence of $H_2O_2$ and water without oxygen [76]. In the presence of oxygen, small amounts of $H_2$ are formed possibly as the result of the water gas shift reaction [76]. Both of these results strongly suggest that OH groups are formed and directly participate in the oxidation mechanism [76]. Similar reactions occur in the electrocatalytic oxidation of CO over Au/C in both acid as well as alkaline media. In order to model the oxidation of CO over bulk Au in solution, we have carried out detailed density functional theory calculations with CO and OH or $H_2O$ adsorbed on a Au(111) surface immersed in water [91]. The resulting activation barrier for the reaction (2) above was calculated to be negligible, whereas the overall reaction energy was calculated to be exothermic by −231 kJ/mol. Various structures along the reaction coordinate are shown in Figure 4. The reaction proceeds via the formation of the OC–OH bond combined with the heterolytic splitting of the O–H bond to form an $H_3O^+$ intermediate which subsequently undergoes a second proton transfer to another water molecule. The overall mechanism for the heterogeneous catalyzed oxidation of CO may proceed via the formation of local electrochemical cell or circuit, where CO is oxidized by OH resulting in the formation of proton and an electron which subsequently catalyzes the reduction of $O_2$ to the active OH intermediates or to $H_2O_2$. This is discussed in somewhat more detail in Section 3.3.2(1). These same ideas on the oxidation of CO can readily be extended to methanol as well as other alcohols.

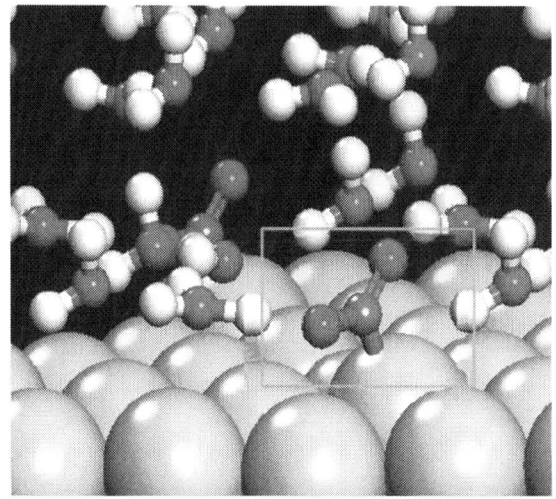

(a)

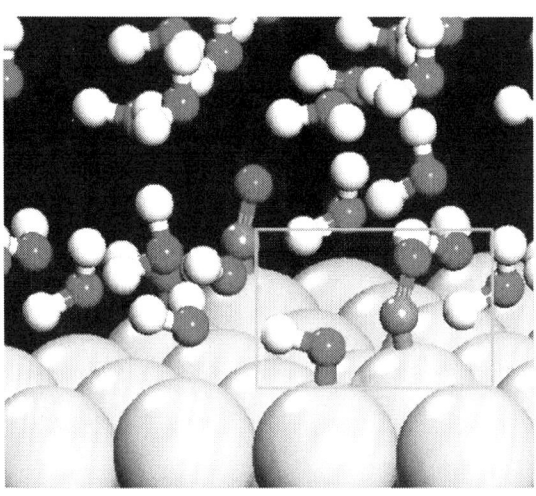

(b)

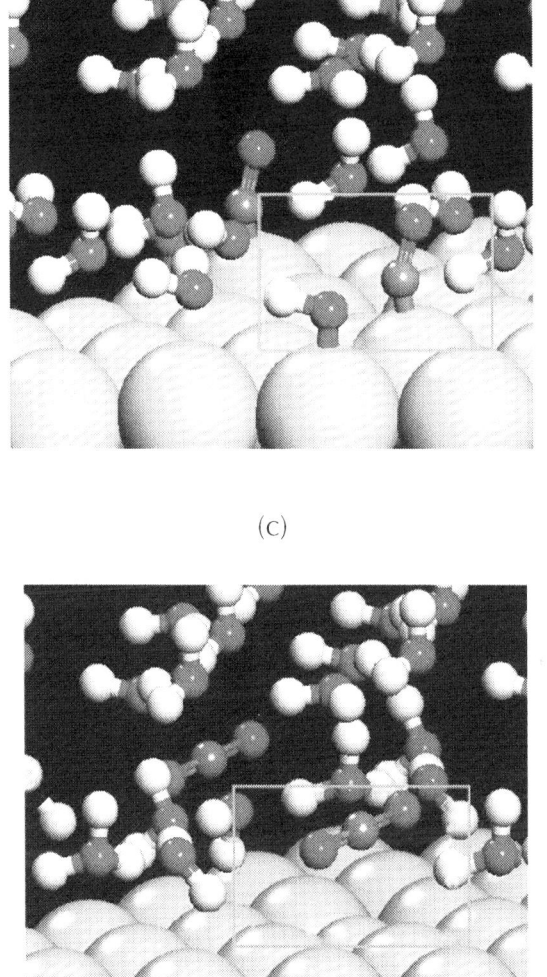

(c)

(d)

**Figure 4:** DFT-calculated reaction coordinate for the oxidation of CO via adsorbed OH* over Au. The reaction proceeds via: (a) the coadsorption of CO* and OH*, (b) the nucleophilic attack of OH* on CO* to form the HO*–*CO surface intermediate, (c) cleavage of O–H bond in the transition state, and (d) formation of $CO_2$ and $H_3O^+$ products. The active CO* and OH* surface species are highlighted in green as they proceed through the reaction.

# Alcohol Oxidation

The increasing demand to shift from petroleum-based fuels and chemical to those derived from biomass has significantly increased efforts in both the electrocatalytic and catalytic oxidation of carbohydrates feedstocks. Significant efforts have been focused on the selective oxidation of methanol, ethanol, glycerol, and other $C_2$–$C_6$ polyols into chemical intermediates via heterogeneous catalysis [92–98] as well as the total oxidation of these fuels to electrical energy via electrocatalysis [99–107]. In the next few sections, we compare some of the similarities and differences involved in the catalytic and electrocatalytic oxidation of these alcohols in acidic as well as alkaline media.

# Alcohol Oxidation in Acidic Media

## *(1) Catalytic*

The catalytic oxidation of glycerol as well as ethanol is rather low when carried out in neutral solution over Pd and Pt with reported turnover frequencies of 0.05 and 0.06 s$^{-1}$, respectively [98]. The products that form over these metals are predominantly intermediate aldehydes and ketones that result from dehydrogenation with only 25% selectivity to form the acid product. Glycerol oxidation, for example, leads to the formation of glyceraldehyde as well as dihydroxyacetone without further oxidation to $C_1$ or $C_2$ acids or $CO_2$. Supported Au clusters under neutral and acidic conditions were found to be completely inactive [98]. The dehydrogenation reactivity that results over Pd and Pt can proceed via C–H and O–H activation over the metal or by adsorbed oxygen or hydroxyl intermediates that form upon $O_2$ dissociation or by the subsequent reaction of adsorbed oxygen with water. DFT-calculated activation barriers suggest that O–H activation of ethanol occurs via the reaction of adsorbed ethanol with OH*, whereas the C–H activation of ethanol preferentially occurs via metal sites on the Pt and Pd surfaces [98]. Both reactions, however, are limited on Pt and Pd, as the high surface coverages on these metals result in barriers for both C–H and O–O activation that are significantly greater than 100 kJ/mol. The oxidation of alcohols over Au does not proceed over Au alone under neutral or acidic conditions as gold cannot activate water, $O_2$, or alcohol [98].

## (2) Electrocatalytic

The oxidation of methanol, ethanol, glycerol, and other alcohols occurs in acidic media but requires significantly higher overpotentials than reactions carried out in alkaline media [79,100, 108–119]. The oxidation of methanol proceeds over most transition metals through a sequence of elementary C–H and O–H bond activation steps which occur on the metal, and ultimately result in the formation of CO [79, 109, 110]. Through the combination of cyclic voltammetry, chronoamperometry, and DFT studies, we demonstrated that the methanol decomposition occurs via a dual path mechanism [1, 120]. At potentials below 0.35 V, the mechanism proceeds predominantly through a sequence of C–H activation steps to form the hydroxyl methylene (CHOH*) intermediate that subsequent breaks the O–H and C–H bonds to form CO*. At potentials above 0.35 V, the O–H bond of methanol can also be activated, resulting in the formation of formaldehyde which can desorb or continue on to form CO. The onset potential was found to be a function of the Pt surface structure. The dual paths for the oxidation of methanol to CO over Pt(111) along with their corresponding reaction energies calculated at 0.5 V RHE are shown in Figure 5 [1, 120]. While CO is an intermediate to $CO_2$, it readily builds up on the surface and poisons more active metals such as Pt. As such, Pt is typically alloyed with a more oxophilic metal such as Ru to promote the adsorption and dissociation of water thus creating bifunctional sites on the surface [79, 109,110]. The OH groups that result interrupt the CO adlayer and readily oxidize CO to $CO_2$. The addition of Ru also helps to weaken the Pt–CO bond thus enhancing CO desorption. The addition of Ru to Pt lowers the overpotential for CO oxidation by ~0.25 V [80–82, 121] as it prevents CO poisoning. The oxidation of ethanol, glycerol and other larger alcohols in acidic media result in dehydrogenation which forms the corresponding aldehyde [114–117]. The subsequent activation of the C–C bond, however, is very difficult over typical metals such as Pt, Pd, or Au, and as such, very limited $CO_2$ is formed. The dehydrogenation routes are identical to those presented above for the catalytic oxidation of the same alcohols. The higher potentials used electrochemically can activate water at higher potentials, and thus result in the formation of surface hydroxyl intermediates that can subsequently oxidize the aldehyde and ketone intermediates [115]. This leads to the formation of acids as well as carboxylate intermediates which inhibit the surface under acidic conditions and prevent the formation of $CH_x$ intermediates on closed-packed crystal surfaces. Feliu et al. have shown that steps on specific Pt surfaces Pt(554) and Pt(110) can begin to enhance C–C bond breaking and $CO_2$ formation, but rate is still very limited [111, 114]. The electrooxidation of ethanol over Pt in acidic media has two major limitations which prevent its viability as was discussed by Lia et al. [115]. The first relates to the fact that reaction predominantly produces acetate and acetic acid intermediates, thus resulting in only 2 and 4 electrons, respectively, which are only very minor contributions to total possible current. Both are thus unwanted side products for fuel cell applications. The second limitation is that the path to $CO_2$ is rather difficult in that it requires the activation of C–C bond as well as the oxidation of both the $CH_x$ and CO intermediates that form. Both of these intermediates tend to inhibit or poison metal surfaces at lower potentials [48]. The characteristic difference between the catalytic oxidation and electrocatalytic oxidation of alcohols in acidic media lies in the generation of the active surface intermediate. In heterogeneous catalysis, the metal plays an important role in activating oxygen. The surface oxygen can directly activate the alcohol or adsorbed water to generate hydroxyl intermediates that aid in activating and oxidizing the alcohol. Under electrocatalytic conditions, the O–H and C–H bonds of the alcohol can be activated on the metal directly as a result of higher potentials or via adsorbed hydroxyl intermediates that form by the activation of water.

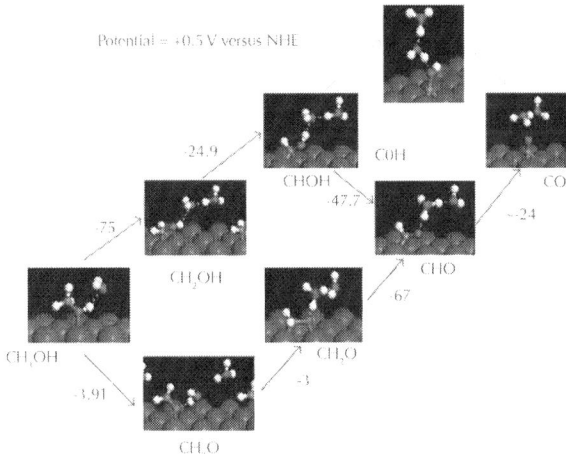

**Figure 5:** DFT-calculated potential-dependent reaction paths for the oxidation of methanol to CO over Pt(111) [1]. At 0.5 V, both experiments and theory point to the onset of dual paths. The primary path, available over a wide range potentials, is shown in red. It proceeds via a sequence of C–H bond activation steps ultimately forming the hydroxyl methylene intermediate (CHOH) before activating the final O–H bond to form CO. The minor path begins between 0.4–0.5 V RHE proceeds via the initial activation of O–H bond of methanol to form a surface methoxy intermediate which subsequently reacts to form formaldehyde consistent with previous speculations [122].

## *Alcohol Oxidation in Alkaline Media*

*(1) Catalytic*

The catalytic oxidation of polyols over Pd, Pt, and Au is much more favorable when carried out in alkaline media [78, 98, 123]. The TOF increases by over an order of magnitude on Pt and over two orders of magnitude on Pd upon the addition of a 2/1 ratio of NaOH to glycerol [98]. Remarkably, the TOF over gold in base is over $6 \, s^{-1}$, whereas the rate over Au in acidic media is negligible [98]. In addition to these increases in TOF, there are also significant improvements in the overall selectivity to form the acid from the alcohol. For glycerol, the selectivity to glyceric acid was found to be 83% over Pd, 70%–78%

over Pt, and 67% over Au [98]. The predominant side products are glycolic and tartronic acid which are shown in the paths outlined in Figure 6. In the presence of base, the second terminal hydroxyl group can be oxidized to form tartronic acid, but this only occurs over Pt and Pd. The oxidation over Au results only in the monofunctional glyceric acid product. In addition to the selective oxygen addition, Pd and Pt can also promote C–C activation resulting in the formation of glycolic, oxalic, lactic, formic, and acetic acids. These C–C activation paths appear to coincide with the formation of hydrogen peroxide [123]. The activation of the C–C bonds have been speculated to occur either through a retroaldol reaction which would be catalyzed by the OH⁻ base, or via oxidation catalyzed by the hydrogen peroxide that forms [78, 123]. Through detailed labeling studies along with DFT simulations, we established a plausible mechanism for the oxidation of alcohols over Au in basic media [98]. The mechanism also helps to explain the unique promotional effects of OH on Au. We discuss here the energetics involved in the reaction over the model Au(111) surface. The first step involves the dehydrogenation of the alcohol to form the corresponding aldehyde. This can proceed via the activation of the O–H and C–H bonds of the alcohol by the metal, adsorbed oxygen, or adsorbed OH intermediates. As one might expect, Au atoms alone cannot activate the O–H bond of the alcohol. The calculated barrier to activate ethanol to ethoxy over Au(111) in the presence of solution was calculated to be 204 kJ/mol (the transition state is shown in Figure 7(a)). The O–H bonds are much more readily activated by the weakly adsorbed OH⁻ intermediates via a mechanism which involves a proton abstraction by the OH⁻ surface intermediate (the transition state is shown in Figure 7(b)). The inability of Au(111) to activate the O–H and C–H bonds is well established, as bulk Au is quite noble. The binding energy of OH$^{*-}$ on Au(111) in water is only −216 kJ/mol (versus −274 kJ/mol on Pt(111)), thus making it quite basic. The weak interaction promotes its ability to readily abstract a proton from a neighboring O–H on the alcohol. Similarly, the activation of the C–H bond of the ethoxy intermediate to form acetaldehyde does not occur over Au(111) alone (see the transition state in Figure 7(c)) but instead proceeds by the reaction of OH$^{*-}$ with the adsorbed ethoxy intermediate with a barrier of only 12 kJ/mol (see the transition state in Figure 7(d)). The aldehyde that results is a key reaction intermediate to form the acid in alkaline media over Au. Most of the initial studies in the aqueous phase catalytic oxidation of alcohols assumed that $O_2$ was responsible for carrying out the oxidation. By carrying out the reaction with $^{18}O$-labeled $O_2$ and $H_2O$, however, we showed that only labeled water found its way into the resulting acid product that forms [98]. DFT results for Au in an aqueous medium showed that this reaction proceeds via a classic nucleophilic attack of $OH^{*-}$ on the adsorbed aldehyde quite similar to thought found for the oxidation of CO. The barrier to form the geminal diol intermediate shown in Figure 8(a) was only 5 kJ/mol when carried out over Au and 42 kJ/mol for reactions in the solution phase Figure 8(b). The final C–H activation of the germinal diol to form the acid can proceed over the Au itself (21 kJ/mol) or via reaction with adsorbed $OH^{*-}$ (29 kJ/mol). A schematic representation of the mechanism which involves the unique reactivity of $OH^-$ intermediates at the aqueous/Au interface for the selective oxidation of the alcohol to the acid is depicted in Figure 9, whereas the corresponding potential energy surface is shown in Figure 10. This oxidation path proceeds without the activation or incorporation of oxygen from $O_2$. While $O_2$ is not directly involved in any of the steps depicted in the mechanism for alcohol oxidation shown in Figure 10, it is critical as the reaction does not occur without it. Oxygen must somehow be intimately coupled with the overall catalytic process which requires the balance of charge and the regeneration of $OH^-$. As the result of theoretical calculations, we showed that oxygen is necessary to remove the electrons that are generated as result of the oxidation of the alcohol. The measured rate for the oxidation reaction is 6.1 turnovers per second per site [98]. As the steps outlined in the cycle presented in Figure 10 consume 4 hydroxyl ions, they generate 4 electrons per every turnover.

$$RCH_2OH + 4OH^- \rightarrow RCOOH + 3H_2O + 4e^- \qquad (3)$$

In order for the reaction to occur catalytically, the electrons that are produced per turnover must be consumed. Each oxygen molecule can effectively remove 4 electrons via the oxygen reduction reaction (ORR).

$$O_2 + 2H_2O + 4e^- \rightarrow 4OH^- \qquad (4)$$

The oxygen reduction reaction is known to occur quite readily over single crystal Au electrodes. DFT calculations were used to determine the reaction energies and the activation barriers for the most relevant steps for ORR. The results reported in Table 1 indicate that the direct dissociation of $O_2$ (reaction 5) does not take place over Au as is well established experimentally. Oxygen instead is reduced by the protons from water and the electrons in the metal to form a peroxo intermediate along with $OH^-$ (reaction 6). The peroxo intermediate is subsequently reduced to hydrogen peroxide (reaction 7) which can subsequently dissociate and result in the formation $2OH^-$ (aq). The intermediate formation of hydrogen peroxide is consistent with observed experimental results and suggestions that peroxide is responsible for C–C bond breaking and formation of shorter acids. This overall ORR cycle removes the 4 electrons from the metal produced via the oxidation, and thus allows the reaction to continue catalytically. The cycle also regenerates $OH^-$(aq). The overall catalytic reaction then involves the direct coupling of both alcohol oxidation and oxygen reduction cycles in one system. This can thus be considered a local, short-circuited, electrochemical cell, where the oxidation and reduction occur simultaneously at the metal/surface interface. The role of $O_2$ then is to simply to remove the electrons from the metal at a rate fast enough to maintain the optimal surface potential of the local electrochemical cell.

**Table 1:** DFT-calculated reaction energies and activation barriers for possible steps involved in the reduction of $O_2$ to OH over Au(111) in aqueous media [98]

|  | $\Delta E_{rxn}$ (kJ/mol) | $\Delta E *$ (kJ/mol) | Reaction |
|---|---|---|---|
| $O_2*+* \rightarrow O*+ O*$ | 41 | 105 | (5) |
| $O_2*+ H_2O* \rightarrow OOH*+ OH*$ | −4 | 16 | (6) |
| $OOH*+* \rightarrow OH*+ O*$ | −56 | 83 | (7) |
| $OOH*+ H_2O* \rightarrow HOOH*+ OH*$ | 37 | 48 | (8) |
| $HOOH*+* \rightarrow OH*+ OH*$ | −86 | 71 | (9) |

**Figure 6:** Reaction pathways speculated in the catalytic and electrocatalytic oxidation of glycerol over Au and Pt. Arrows in red depict paths observed over Au and Pt whereas the paths in black are only observed over Pt. Adapted from [123].

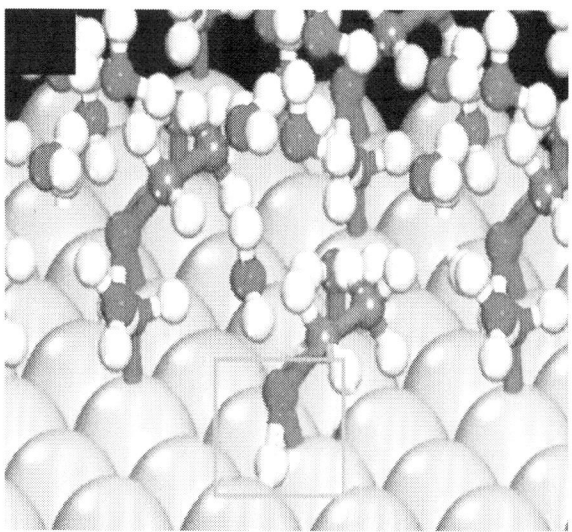

(a)

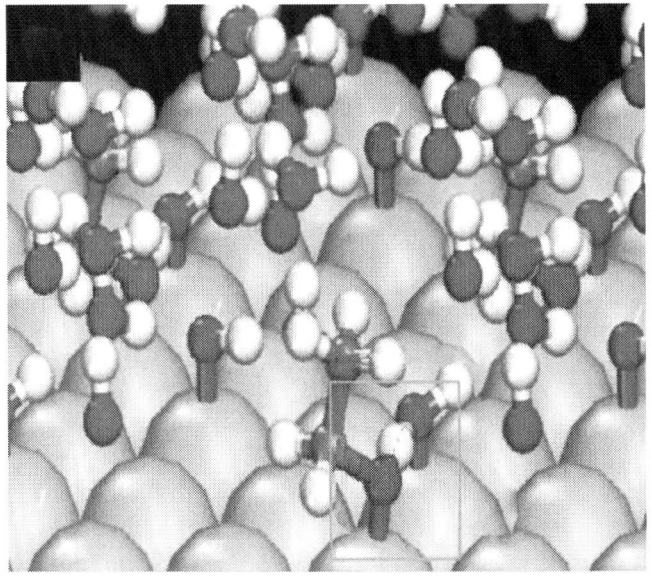

(b)

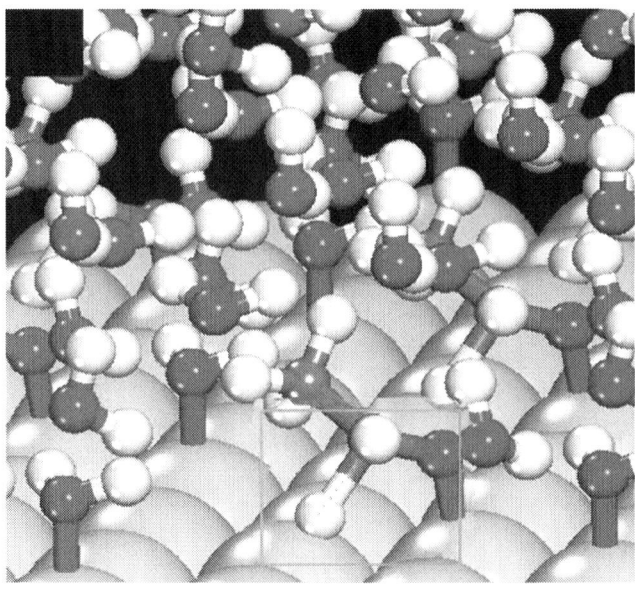

(c)

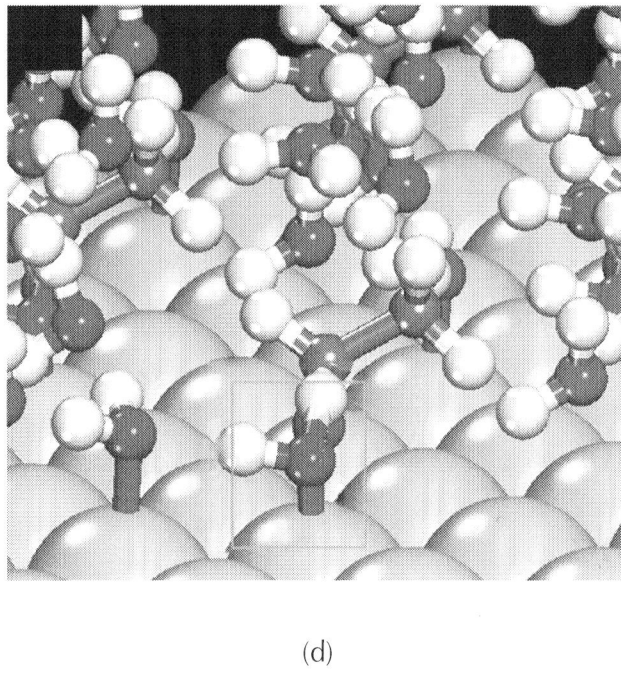

(d)

**Figure 7:** DFT-calculated transition states for the activation of the O–H bond of ethanol adsorbed to Au by (a) metal surface atom or (b) a coadsorbed hydroxyl intermediate on Au(111). The activation of the C–H bonds of the adsorbed ethoxy are activated similarly via (c) Au atoms in the Au(111) surface or (d) via the bound OH groups on the Au(111) surface. The active bond breaking and forming sites are highlighted in green.

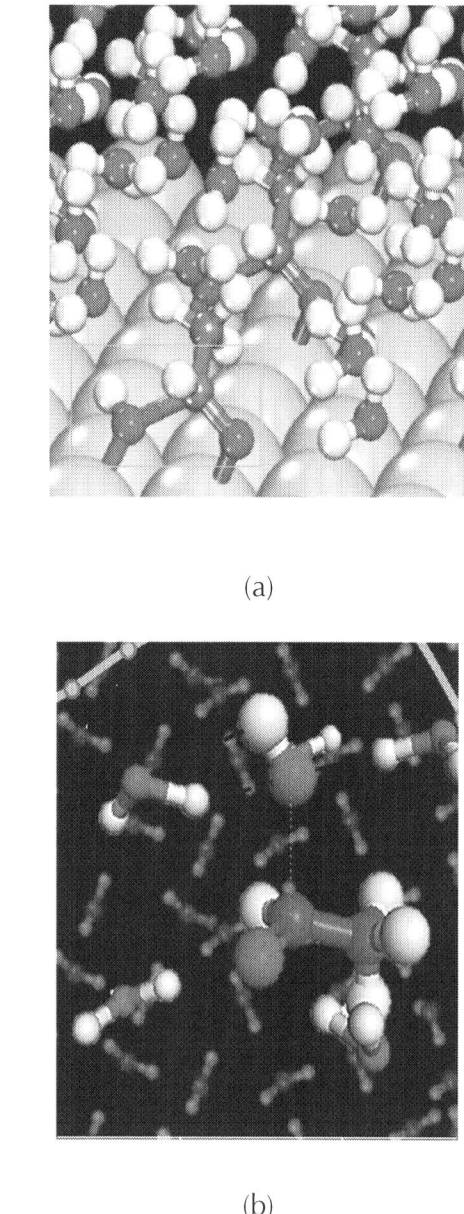

(a)

(b)

**Figure 8**: DFT-calculated transition states and their activation barriers for the oxidation of acetaldehyde on (a) the Au(111) surface or (b) in the presence of solution. The active bond breaking and forming sites are highlighted in green.

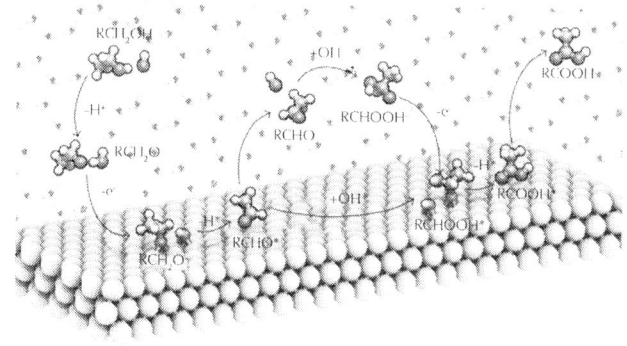

**Figure 9:** Schematic representaion of the reaction paths for ethanol and other polyol oxidation to acids over Au and Pt in alkaline media based on DFT results and isotopic kinetic labeling experiments. (Copyright Science [98]).

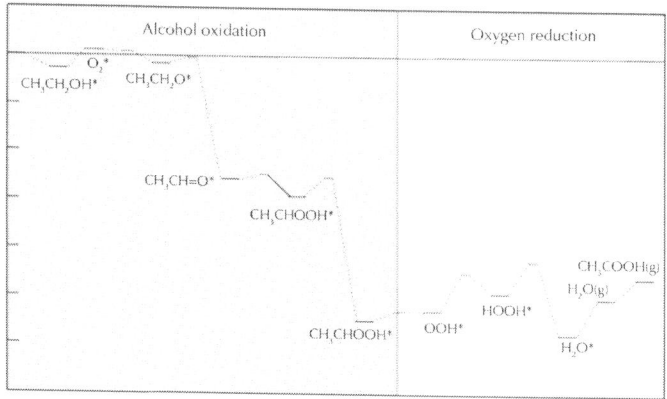

**Figure 10:** DFT-calculated potential energy surface for the catalytic oxidation of ethanol to acetic acid over Au(111) in the presence of base (OH⁻) and in water. Both the OH⁻ as well as the metal are involved in the mechanism. In order establish an overall cycle, $O_2$ is reduced via the electrons that from as a result of the oxidation of the alcohol as well as by the reaction protons that form upon the activation of water. Red lines refer to transition states, whereas the black lines refer to reaction intermediates.

# (2) Electrocatalytic

The development and application of carbonate as well as anion-exchange membrane electrolytes have significantly renewed interest in the development of alkaline-based direct alcohol fuel cells [100]. Many of the reaction intermediates, products, and paths discussed above for the catalytic oxidation of alcohols in alkaline media have also been identified or speculated to take part in the electrocatalytic oxidation of these same alcohols. As such, the mechanistic insights established from catalytic oxidation should be important in understanding the chemistry and the mechanisms that control electrocatalytic oxidation. Similarly, the detailed insights established from electrochemical methods such as cyclic voltammetry should provide new insights into heterogeneous catalytic oxidation. The oxidation of methanol, ethanol, as well as other polyols in alkaline media is thought to proceed either by $4e^-$ or $6e^-$ processes thus resulting in the formation of either formate and carbonate intermediates by reactions (5), or (6) [100]

$$RCH_2OH+5OH^-\rightarrow HCOO^-+4H_2O+4e^- \tag{5}$$

$$RCH2OH+8OH^-\rightarrow CO_3^{2-}+6H_2O+6e^- \tag{6}$$

For methanol, the later reaction is speculated to occur through the formation of CO. In the presence of base, surface $OH^-$ groups readily catalyze the oxidation of methanol to $CO_2$ or carbonate as shown in reactions (5) and (6), respectively. At higher potentials, however, OH binds very strongly to the metal surface and inhibits the reaction. The selectivity to formate or carbonate depends upon the metal, metal surface structure, potential pH, and alcohol concentration present during the reaction. We discuss here the difference in the electrocatalytic oxidation over Pt and Au and compare the changes that result upon changing the pH and concentration. While the oxidation of methanol as well as other alcohols can be carried out in acid media as was discussed above, the activity is rather low. The rates and currents increase significantly with increasing the pH of solution. The electrocatalytic oxidation of methanol as well as other alcohols over Au is quite high in alkaline media [100, 101, 104–108, 110, 118, 124–126]. The higher rates and current densities found in alkaline media are the result of the high reactivity of $OH^-$ anions that are weakly bound to the Au substrate at moderate potentials. The weakly bound hydroxyls are quite basic and will readily activate C–H and O–H bonds (on adsorbed species) as was presented above in the catalytic oxidation of alcohols over Au in basic media. The electrocatalytic activity over different Au substrates including polycrystalline Au, Au(111), and Au(210) all show increases in activity with increases in the solution pH [101]. The adsorption of OH⁻ anions onto Au is thought to significantly enhance the reactivity of the surface over a range of potentials. Electrocatalytic oxidation can, therefore, proceed at much lower potentials where Au is not oxidized. The anodic oxidation of methanol begins to occur with the adsorption of OH⁻. In general, there are two very different regimes that result for methanol oxidation over Au [101]. In the first regime, methanol is actively oxidized by adsorbed OH⁻ intermediates on Au at potentials as low as 0.6 V RHE. The second regime which occurs at much more positive potentials and results in the formation of Au–O monolayers which are much less active for carrying out oxidation. Chen and Lipkowski [127] and Yan et al. [107] provided more details in terms of the changes that take place at the surface as a result of these changes in potential and suggest that there are actually three regimes that result upon increasing the potential. The initial oxidation of methanol and other alcohols proceeds quite readily from 0.07–0.13 V versus Hg/HgO. The current increases linearly with the adsorption of OH⁻ in this regime. There appears to be a weak amount of charge transfer between the OH⁻ and the metal surface. The second regime appears between 0.13–0.3 V versus Hg/HgO and results in a higher degree of charge transfer of OH and the Au sites in the surface. Regime 3 which appears at potentials greater than 0.3 V Hg/HgO is the inactive monolayer Au oxide that results. The oxidation of methanol is thought to proceed via the weakly held OH⁻ intermediates found in both regimes 1 and 2. The more strongly held oxygen intermediates which begin to form in regime 2 are less active. Regime 3 leads to the formation of monolayer Au oxide coverages and thus the loss of the active OH⁻ sites. For methanol oxidation, the reaction is 0.75 order in methanol and 0.55 order in OH [107]. There is a clear negative shift in the potential and an increase in the current density with an increase in the concentration of weakly adsorbed OH⁻ and methanol coverage.

## *Ethanol Oxidation over Au*

Ethanol oxidation is similar to methanol oxidation in that there is a significant shift to lower potentials and increased current that results upon increasing pH. Ethanol oxidation proceeds at a potential of 0.6 V versus RHE in alkaline solution which is 0.3 V lower than that reported in acid [115]. In addition there is a shift of −0.1 V in the maximum from 1.35 V to 1.25 V. The current density in alkaline media is found to be over an order of magnitude higher than that in neutral or acid conditions [115]. Regardless of the potential or pH there is negligible C–C bond breaking that occurs on Au, and as such, there is very little to no $CO_2$ formation. The mechanism is thought to be very similar to the path outlined in Figure 10 and presented above for the catalytic oxidation of ethanol. The main path involves the oxidation of ethanol to acetaldehyde through the activation of the acidic O–H and C–H bonds by weakly adsorbed hydroxyl intermediates on the Au surface. Acetaldehyde subsequently reacts with weakly held $OH^-$ intermediates via simple nucleophilic attack depending of $OH^-$ on the C=O bond of the aldehyde to form the geminal diol intermediate which can further react with $OH^-$ to form acetic acid or adsorbed acetate in alkaline media.

## Extension to Polyols

These same ideas have been further extended to larger polyols such as glycerol and glucose in alkaline media [104–106, 125]. Glycerol, for example, reacts via a $4e^-$ transfer process to form glyceric acid on both Au as well as Pt electrodes. At higher potentials, the C–C bond of glyceric acid can further oxidize via a two electron transfer step to form glycolic and formic acids [125]. While this occurs over both Au and Pt, the selectivity to the over-oxidized glycolic state is much higher on Au. This was thought to be due to the fact that Pt deactivates by the formation of a surface oxide intermediate at much lower potential (0.9 V) than that found on Au (~1.3 V) [125]. This results in much lower oxidation potentials and a much narrower window for the subsequent oxidation to proceed. Glyceric acid as well as glycolic acid can both undergo further oxidation on Pt to form tartronic as well as oxalic acid, respectively. The subsequent $OH^-$ addition steps, however, only appear to proceed over Pt as neither of the di-substituted products appear over Au. The results reported by Kwon and Koper [125] for the electrocatalytic oxidation of glycerol discussed here are very similar to those reported by Ketchie et al. [78] for the catalytic oxidation of glycerol. Both show that C–C bond breaking steps can readily take place in the presence of

OH⁻ to form glycolic, formic, and oxalic acids on Au as well as on Pt. Both also reveal that Au will only oxidize one of the terminal $-CH_2OH$ bonds of the polyol, whereas Pt can oxidize both. It is very likely that the detailed mechanism for the formation of glyceric acid established for catalytic systems holds also for the electrocatalytic oxidation. The elementary steps involved in C–C bond breaking are not known for either the catalytic or the electrocatalytic paths. The catalytic routes have been speculated to occur via oxidation with hydrogen peroxide or by base catalyzed paths. There is no evidence, however, for the formation of hydrogen peroxide electrocatalytically and more likely the reaction proceeds solely via the reaction with base.

The similarities and differences between the catalytic and electrocatalytic oxidation of alcohols is very informative. A closer analysis between the two suggests that the two are nearly equivalent with the exception that oxidation of the alcohol which occurs at the anode is decoupled from the oxygen reduction which occurs at the cathode. The two electrodes communicate via charge and ion transfer. In the heterogeneous catalytic system, the alcohol is oxidized where the electrons are directly used at the same aqueous metal interface to reduce $O_2$. The rate of oxidation is thus directed by the rate at which $O_2$ can be reduced by electrons. This is directly analogous to that which happens in the PEM alcohol fuel cells.

CO and alcohol oxidation make up just two reaction systems. One can readily draw analogies to other catalytic reaction systems carried out in aqueous media. Desai and Neurock for example carried out first principles DFT calculations to show that an aqueous medium could readily facilitate the catalytic hydrogenation of adsorbed oxygenates [83, 84,128]. They showed that adsorbed hydrogen could undergo an electron transfer coupled with proton transfer at the metal interface to form a local proton in the form of a hydronium ion that could readily transfer through solution and attach itself with the negatively charged adsorbed surface intermediate. The solution phase here acts as a cocatalyst. As such, the solution provides for a short-circuited electrocatalytic cell which allows for electron and proton generation transfer and facile recombination. The ability carry out this heterolytic proton transfer process is controlled by the work function or "electron affinity" of the metal. Metals such as Pd and Pt should have the ability to convert adsorbed hydrogen into protons that reside in solution and electrons which remain in the metal. This was demonstrated by both Wagner and Moylan [129] and Kizhakevariam and Struve [130] who both showed that adsorbed hydrogen on ideal Pt substrates in the presence of water in UHV (at low temperature) forms hydronium ions in solution near the surface. These results are consistent with the results from theory. More generally the results suggest that the presence of protic media along with high work function metals can carry out hydrogenation reactions much more efficiently through proton coupled electron transfer processes that mimic electrocatalysis.

# SUMMARY

While fundamental information concerning reaction mechanisms, active sights and catalytic kinetics gleaned from in situ spectroscopy, detailed theoretical simulations and rigorous kinetic studies for gas phase heterogeneous catalysis has helped guide the development of electrocatalytic systems, the complexity of the reaction environment has often precluded more in-depth or quantitative analyses. The tremendous advances in spectroscopy along with theory, that have taken place over the past few decades, however, have allowed for more detailed resolution of the molecular transformations that occur in electrocatalytic systems along with a more detailed following of the nature of the active centers and their environment. There appears to be an important and growing trend where this knowledge and guiding principles from electrocatalysis are being used to guide heterogeneous the complex aqueous and solvent-based catalytic processes. The knowledge of the complex electrified interface in electrocatalysis bears a number of common similarities to the aqueous/metal interface for catalytic reactions carried out the presence of solution. More detailed fundamental studies which attempt to rigorously compare heterogeneous catalysis in solution with electrocatalysis will continue and will likely be crucial in establishing the links between the two.

# ACKNOWLEDGMENTS

A. Wieckowski gratefully acknowledges support from the National Science Foundation under Grant no. NSF CHE06-51083 A. Wieckowski and ARO under Grant no. W911NF-08-1-0309. M. Neurock gratefully acknowledges support from the Office of Basic Energy Sciences under Award no. ERKCC61 for the work on metal/solution interfaces. (This work is part of the Fluid Interface Reactions, Structures and Transport (FIRST) Center, an Energy Frontier Research Center funded by the U.S. Department of Energy, Office of Science, Office of Basic Energy Sciences), U.S. Department of Energy, Division of Chemical Sciences, Office of Basic Energy Sciences (DE-FG02-07ER15894) for the work on oxygen reduction, the National Science Foundation Research Center for Biorenewable Chemicals (EEC-0813570), and the National Science Foundation PIRE (NSF OISE-0730277) for the work on alcohol oxidation. M. Neurock would also kindly acknowledge the computational time at the Environmental Molecular Science Laboratory, a national scientific user facility sponsored by the Department of Energy's Office of Biological and Environmental Research and located at Pacific Northwest National Laboratory, the National Center for Computational Sciences at Oak Ridge National Laboratory and the National Energy Research Scientific Computing Center, which is supported by the Office of Science of the U.S. Department of Energy which was used to carry out the work. Lastly, the authers would like to thank Professor Robert J. Davis, David Hibbitts, and Craig Plaisance for helpful discussions.

# REFERENCES

1. D. Cao, G. Q. Lu, A. Wieckowski, S. A. Wasileski, and M. Neurock, "Mechanisms of methanol decomposition on platinum: a combined experimental and ab initio approach," Journal of Physical Chemistry B, vol. 109, no. 23, pp. 11622–11633, 2005.

2. M. J. Janik, S. A. Wasileski, C. D. Taylor, and M. Neurock, "First principles simulation of the active sites and reaction environment in electrocatalysis," in Fuel Cell Catalysis: A Surface Science Approach, M. Koper, Ed., John Wiley and Sons, 2008.

3. M. Koper, Ed., Fuel Cell Catalysis: A Surface Science Approach, John Wiley and Sons, 2008.

4. A. Wieckowski and J. K. Nørskov, Fuel Cell Science: Theory, Fundamentals, and Bio-Catalysis, John Wiley and Sons, 2010.

5. R. A. van Santen and M. Neurock, "Theory of surface chemical reactivity," in Handbook of Catalysis, H. K. G. Ertl and J. Weitcamp, Eds., pp. 942–958, Springer, 1997.

6.    R. A. van Santen and M. Neurock, Molecular Heterogeneous Catalysis: A Mechanistic and Computational Approach, VCH-Wiley, 2006.

7.    R. A. van Santen, M. Neurock, and S. G. Shetty, "Reactivity theory of transition-metal surfaces: a brønsted-evans- polanyi linear activation energy-free-energy analysis,"Chemical Reviews, vol. 110, no. 4, pp. 2005–2048, 2010.

8.    T. Bligaard and J. K. Nørskov, "Ligand effects in heterogeneous catalysis and electrochemistry," Electrochimica Acta, vol. 52, no. 18, pp. 5512–5516, 2007

9.    J. Greeley, T. F. Jaramillo, J. Bonde, I. B. Chorkendorff, and J. K. Nørskov, "Computational high-throughput screening of electrocatalytic materials for hydrogen evolution," Nature Materials, vol. 5, no. 11, pp. 909–913, 2006.

10.   P. Liu, A. Logadottir, and J. K. Nørskov, "Modeling the electro-oxidation of CO and $H_2/CO$ on Pt, Ru, PtRu and $Pt_3Sn$," Electrochimica Acta, vol. 48, no. 25-26, pp. 3731–3742, 2003.

11.   J. K. Nørskov, J. Rossmeisl, A. Logadottir et al., "Origin of the overpotential for oxygen reduction at a fuel-cell cathode," Journal of Physical Chemistry B, vol. 108, no. 46, pp. 17886–17892, 2004.

12.   J. Rossmeisl, J. K. Nørskov, C. D. Taylor, M. J. Janik, and M. Neurock, "Calculated phase diagrams for the electrochemical oxidation and reduction of water over Pt(111)," Journal of Physical Chemistry B, vol. 110, no. 43, pp. 21833–21839, 2006.

13.   A. B. Anderson, "$O_2$ reduction and CO oxidation at the Pt-electrolyte interface. The role of $H_2O$ and OH adsorption bond strengths," Electrochimica Acta, vol. 47, no. 22-23, pp. 3759–3763, 2002.

14.   A. B. Anderson and Y. Cai, "Calculation of the Tafel plot for $H_2$ oxidation on Pt(100) from potential-dependent activation energies," Journal of Physical Chemistry B, vol. 108, no. 52, pp. 19917–19920, 2004.

15.   A. B. Anderson, J. Roques, S. Mukerjee, V. S. Murthi, N. M. Markovic, and V. Stamenkovic, "Activation energies for oxygen reduction on platinum alloys: theory and experiment," Journal of Physical Chemistry B, vol. 109, no. 3, pp. 1198–1203, 2005.

16. J. Roques and A. B. Anderson, "Electrode potential-dependent stages in $OH_{ads}$ formation on the $Pt_3Cr$ alloy (111) surface," Journal of the Electrochemical Society, vol. 151, no. 11, pp. E340–E347, 2004.

17. J. Roques, A. B. Anderson, V. S. Murthi, and S. Mukerjee, "Potential shift for OH(ads) formation on the Pt skin on $Pt_3Co(111)$ electrodes in acid theory and experiment," Journal of the Electrochemical Society, vol. 152, no. 6, pp. E193–E199, 2005.

18. R. A. Sidik and A. B. Anderson, "Density functional theory study of $O_2$ electroreduction when bonded to a Pt dual site," Journal of Electroanalytical Chemistry, vol. 528, no. 1-2, pp. 69–76, 2002.

19. A. Lamm, H. Gasteiger, and W. Vielstich, Eds., Fuel Cell Handbook, Vol. 2, Electrocatalysis, Wiley-VCH, 2003.

20. T. D. Jarvi and E. M. Stuve, "Fundamental aspects of vacuum and electrocatalytic reactions of methanol and formic acid on platinum surfaces," in Electrocatalysis, J. Lipkowski and P. N. Ross, Eds., Wiley-VCH, 1998.

21. G. Q. Lu, A. Lagutchev, D. D. Dlott, and A. Wieckowski, "Quantitative vibrational sum-frequency generation spectroscopy of thin layer electrochemistry: CO on a Pt electrode," Surface Science, vol. 585, no. 1-2, pp. 3–16, 2005.

22. P. K. Babu, H. S. Kim, S. T. Kuk et al., "Activation of nanoparticle Pt-Ru fuel cell catalysts by heat treatment: A195Pt NMR and electrochemical study," Journal of Physical Chemistry B, vol. 109, no. 36, pp. 17192–17196, 2005.

23. H. Yano, J. Inukai, H. Uchida et al., "Particle-size effect of nanoscale platinum catalysts in oxygen reduction reaction: an electrochemical and $_{195}$Pt EC-NMR study," Physical Chemistry Chemical Physics, vol. 8, no. 42, pp. 4932–4939, 2006.

24. T. Kobayashi, P. K. Babu, J. H. Chung, E. Oldfield, and A. Wieckowski, "Coverage dependence of CO surface diffusion on Pt nanoparticles: an EC-NMR study," Journal of Physical Chemistry C, vol. 111, no. 19, pp. 7078–7083, 2007.

25. S. Maniguet, R. J. Mathew, and A. E. Russell, "EXAFS of carbon monoxide oxidation on supported Pt fuel cell electrocatalysts," Journal of Physical Chemistry B, vol. 104, no. 9, pp. 1998–2004, 2000.

26.  J. McBreen and S. Mukerjee, "In situ X-ray absorption studies of a Pt-Ru electrocatalyst," Journal of the Electrochemical Society, vol. 142, no. 10, pp. 3399–3404, 1995.

27.  F. J. Scott, C. Roth, and D. E. Ramaker, "Kinetics of CO poisoning in simulated reformate and effect of Ru island morphology on PtRu fuel cell catalysts as determined by operando X-ray absorption near edge spectroscopy," Journal of Physical Chemistry C, vol. 111, no. 30, pp. 11403–11413, 2007.

28.  F. J. Scott, S. Mukerjee, and D. E. Ramaker, "CO coverage/oxidation correlated with PtRu electrocatalyst particle morphology in 0.3 M methanol by in situ XAS," Journal of the Electrochemical Society, vol. 154, no. 5, pp. A396–A406, 2007.

29.  C. Roth, N. Benker, T. Buhrmester et al., "Determination of O[H] and CO coverage and adsorption sites on PtRu electrodes in an operating PEM fuel cell," Journal of the American Chemical Society, vol. 127, no. 42, pp. 14607–14615, 2005.

30.  G. Samjeske, A. Miki, S. Ye, and M. Osawa, "Mechanistic study of electrocatalytic oxidation of formic acid at platinum in acidic solution by time-resolved surface-enhanced infrared absorption spectroscopy," Journal of Physical Chemistry B, vol. 110, no. 33, pp. 16559–16566, 2006.

31.  G. Samjeske, A. Miki, and M. Osawa, "Electrocatalytic oxidation of formaldehyde on platinum under galvanostatic and potential sweep conditions studied by time-resolved surface-enhanced infrared spectroscopy," Journal of Physical Chemistry C, vol. 111, no. 41, pp. 15074–15083, 2007.

32.  A. B. Anderson, "Theory at the electrochemical interface: reversible potentials and potential-dependent activation energies," Electrochimica Acta, vol. 48, no. 25-26, pp. 3743–3749, 2003.

33.  A. B. Anderson, Y. Cai, R. A. Sidik, and D. B. Kang, "Advancements in the local reaction center electron transfer theory and the transition state structure in the first step of oxygen reduction over platinum," Journal of Electroanalytical Chemistry, vol. 580, no. 1, pp. 17–22, 2005.

34.  A. B. Anderson, N. M. Neshev, R. A. Sidik, and P. Shiller, "Mechanism for the electrooxidation of water to OH and O

bonded to platinum: quantum chemical theory," Electrochimica Acta, vol. 47, no. 18, pp. 2999–3008, 2002.

35. Y. Cai and A. B. Anderson, "The reversible hydrogen electrode: potential-dependent activation energies over platinum from quantum theory," Journal of Physical Chemistry B, vol. 108, no. 28, pp. 9829–9833, 2004.

36. E. Santos, A. Lundin, K. Potting, P. Quaino, and W. Schmickler, "Model for the electrocatalysis of hydrogen evolution," Physical Review B, vol. 79, no. 23, Article ID 235436, 2009.

37. E. Santos and W. Schmickler, "Electronic interactions decreasing the activation barrier for the hydrogen electro-oxidation reaction," Electrochimica Acta, vol. 53, no. 21, pp. 6149–6156, 2008.

38. F. Wilhelm, W. Schmickler, R. R. Nazmutdinov, and E. Spohr, "A model for proton transfer to metal electrodes," Journal of Physical Chemistry C, vol. 112, no. 29, pp. 10814–10826, 2008.

39. J. S. Filhol and M. Neurock, "Elucidation of the electrochemical activation of water over Pd by first principles," Angewandte Chemie, vol. 45, no. 3, pp. 403–406, 2006.

40. C. Taylor, R. G. Kelly, and M. Neurock, "First-principles calculations of the electrochemical reactions of water at an immersed Ni(111)/$H_2O$ interface," Journal of the Electrochemical Society, vol. 153, no. 12, pp. E207–E214, 2006.

41. C. D. Taylor and M. Neurock, "Theoretical insights into the structure and reactivity of the aqueous/metal interface," Current Opinion in Solid State and Materials Science, vol. 9, no. 1-2, pp. 49–65, 2005.

42. C. D. Taylor, M. Neurock, and J. R. Scully, "First-principles investigation of the fundamental corrosion properties of a model $Cu_{38}$ nanoparticle and the (111), (113) surfaces," Journal of the Electrochemical Society, vol. 155, no. 8, pp. C407–C414, 2008.

43. C. D. Taylor, S. A. Wasileski, J. S. Filhol, and M. Neurock, "First principles reaction modeling of the electrochemical interface: consideration and calculation of a tunable surface potential from atomic and electronic structure," Physical Review B, vol. 73, no. 16, Article ID 165402, pp. 1–16, 2006.

44. M. Otani, I. Hamada, O. Sugino, Y. Morikawa, Y. Okamoto, and T. Ikeshoji, "Structure of the water/platinum interface—a first

principles simulation under bias potential,"Physical Chemistry Chemical Physics, vol. 10, no. 25, pp. 3609–3612, 2008.

45. M. Otani, I. Hamada, O. Sugino, Y. Morikawa, Y. Okamoto, and T. Ikeshoji, "Electrode dynamics from first principles," Journal of the Physical Society of Japan, vol. 77, no. 2, Article ID 024802, 2008.

46. R. Jinnouchi and A. B. Anderson, "Electronic structure calculations of liquid-solid interfaces: combination of density functional theory and modified Poisson-Boltzmann theory," Physical Review B, vol. 77, no. 24, Article ID 245417, 2008.

47. J. Rossmeisl, E. Skulason, M. E. Bjorketun, V. Tripkovic, and J. K. Nørskov, "Modeling the electrified solid-liquid interface," Chemical Physics Letters, vol. 466, no. 1–3, pp. 68–71, 2008.

48. M. Neurock, W. Vielstich, H. A. Gasteiger, and H. Yokokawa, "First principles modeling for the electrooxidation of small molecules," in Handbook of Fuel Cells, W. Vielstich, H. A. Gasteiger, and H. Yokokawa, Eds., John Wiley and Sons, 2009.

49. G. Kresse and J. Furthmuller, "Efficiency of ab-initio total energy calculations for metals and semiconductors using a plane-wave basis set," Computational Materials Science, vol. 6, no. 1, pp. 15–50, 1996.

50. G. Kresse and J. Furthmuller, "Efficient iterative schemes for ab initio total-energy calculations using a plane-wave basis set," Physical Review B, vol. 54, no. 16, pp. 11169–11186, 1996.

51. G. Henkelman and H. Jónsson, "Improved tangent estimate in the nudged elastic band method for finding minimum energy paths and saddle points," Journal of Chemical Physics, vol. 113, no. 22, pp. 9978–9985, 2000.

52. G. Henkelman, B. P. Uberuaga, and H. Jónsson, "Climbing image nudged elastic band method for finding saddle points and minimum energy paths," Journal of Chemical Physics, vol. 113, no. 22, pp. 9901–9904, 2000.

53. G. Henkelman and H. Jónsson, "A dimer method for finding saddle points on high dimensional potential surfaces using only first derivatives," Journal of Chemical Physics, vol. 111, no. 15, pp. 7010–7022, 1999.

54.  R. R. Davda, J. W. Shabaker, G. W. Huber, R. D. Cortright, and J. A. Dumesic, "Aqueous-phase reforming of ethylene glycol on silica-supported metal catalysts,"Applied Catalysis B, vol. 43, no. 1, pp. 13–26, 2003.

55.  P. Gallezot, N. Nicolaus, G. Fleche, and A. Perrard, "Glucose hydrogenation on ruthenium catalysts in a trickle-bed reactor," Journal of Catalysis, vol. 180, no. 1, pp. 51–55, 1998.

56.  D. K. Sohounloue, C. Montassier, and J. Barbier, "Catalytic hydrogenolysis of sorbitol," Reaction Kinetics and Catalysis Letters, vol. 22, no. 3-4, pp. 391–397, 1983.

57.  M. B. Valenzuela, C. W. Jones, and P. K. Agrawal, "Batch aqueous-phase reforming of woody biomass," Energy and Fuels, vol. 20, no. 4, pp. 1744–1752, 2006.

58.  N. Dimitratos, C. Messi, F. Porta, L. Prati, and A. Villa, "Investigation on the behaviour of Pt(0)/carbon and Pt(0),Au(0)/carbon catalysts employed in the oxidation of glycerol with molecular oxygen in water," Journal of Molecular Catalysis A, vol. 256, no. 1-2, pp. 21–28, 2006.

59.  R. D. Cortright, R. R. Davda, and J. A. Dumesic, "Hydrogen from catalytic reforming of biomass-derived hydrocarbons in liquid water," Nature, vol. 418, no. 6901, pp. 964–967, 2002.

60.  J. W. Shabaker, G. W. Huber, R. R. Davda, R. D. Cortright, and J. A. Dumesic, "Aqueous-phase reforming of ethylene glycol over supported platinum catalysts,"Catalysis Letters, vol. 88, no. 1-2, pp. 1–8, 2003.

61.  E. P. Maris, W. C. Ketchie, V. Oleshko, and R. J. Davis, "Metal particle growth during glucose hydrogenation over Ru/SiO$_2$ evaluated by X-ray absorption spectroscopy and electron microscopy," Journal of Physical Chemistry B, vol. 110, no. 15, pp. 7869–7876, 2006.

62.  E. P. Maris, W. C. Ketchie, M. Murayama, and R. J. Davis, "Glycerol hydrogenolysis on carbon-supported PtRu and AuRu bimetallic catalysts," Journal of Catalysis, vol. 251, no. 2, pp. 281–294, 2007.

63.  W. C. Ketchie, E. P. Maris, and R. J. Davis, "In-situ X-ray absorption spectroscopy of supported Ru catalysts in the aqueous phase," Chemistry of Materials, vol. 19, no. 14, pp. 3406–3411, 2007.

64. M. Haruta, T. Kobayashi, H. Sano, and N. Yamada, "Novel gold catalysts for the oxidation of carbon monoxide at a temperature far below 0.DEG.C," Chemistry Letters, vol. 16, no. 2, pp. 405–408, 1987.

65. M. Haruta, N. Yamada, T. Kobayashi, and S. Iijima, "Gold catalysts prepared by coprecipitation for low-temperature oxidation of hydrogen and of carbon monoxide," Journal of Catalysis, vol. 115, no. 2, pp. 301–309, 1989.

66. G. C. Bond, C. Louis, and D. T. Thompson, Catalysis by Gold, Imperial College Press, London, UK, 2006.

67. R. Meyer, C. Lemire, S. K. Shaikhutdinov, and H. Freund, "Surface chemistry of catalysis by gold," Gold Bulletin, vol. 37, no. 1-2, pp. 72–124, 2004.

68. M. Valden, X. Lai, and D. W. Goodman, "Onset of catalytic activity of gold clusters on titania with the appearance of nonmetallic properties," Science, vol. 281, no. 5383, pp. 1647–1650, 1998.

69. N. Lopez, T. V. W. Janssens, B. S. Clausen et al., "On the origin of the catalytic activity of gold nanoparticles for low-temperature CO oxidation," Journal of Catalysis, vol. 223, no. 1, pp. 232–235, 2004.

70. Q. Fu, H. Saltsburg, and M. Flytzani-Stephanopoulos, "Active nonmetallic Au and Pt species on ceria-based water-gas shift catalysts," Science, vol. 301, no. 5635, pp. 935–938, 2003.

71. J. Guzman and B. C. Gates, "Simultaneous presence of cationic and reduced gold in functioning MgO-supported CO oxidation catalysts: evidence from X-ray absorption spectroscopy," Journal of Physical Chemistry B, vol. 106, no. 31, pp. 7659–7665, 2002.

72. L. M. Molina and B. Hammer, "Active role of oxide support during CO oxidation at Au/MgO," Physical Review Letters, vol. 90, no. 20, 4 pages, 2003.

73. C. K. Costello, M. C. Kung, H. S. Oh, Y. Wang, and H. H. Kung, "Nature of the active site for CO oxidation on highly active Au/ -Al$_2$O$_3$," Applied Catalysis A, vol. 232, no. 1-2, pp. 159–168, 2002.

74. H. H. Kung, M. C. Kung, and C. K. Costello, "Supported Au catalysts for low temperature CO oxidation," Journal of Catalysis, vol. 216, no. 1-2, pp. 425–432, 2003.

75. W. B. Kim, T. Voitl, G. J. Rodriguez-Rivera, and J. A. Dumesic, "Powering fuel cells with CO via aqueous polyoxometalates and gold catalysts," Science, vol. 305, no. 5688, pp. 1280–1283, 2004.

76. M. A. Sanchez-Castillo, C. Couto, W. B. Kim, and J. A. Dumesic, "Gold-nanotube membranes for the oxidation of CO at gas-water interfaces," Angewandte Chemie, vol. 43, no. 9, pp. 1140–1142, 2004.

77. W. C. Ketchie, Y. L. Fang, M. S. Wong, M. Murayama, and R. J. Davis, "Influence of gold particle size on the aqueous-phase oxidation of carbon monoxide and glycerol,"Journal of Catalysis, vol. 250, no. 1, pp. 94–101, 2007.

78. W. C. Ketchie, M. Murayama, and R. J. Davis, "Promotional effect of hydroxyl on the aqueous phase oxidation of carbon monoxide and glycerol over supported Au catalysts," Topics in Catalysis, vol. 44, no. 1-2, pp. 307–317, 2007.

79. N. M. Markoví and P. N. Ross, "Surface science studies of model fuel cell electrocatalysts," Surface Science Reports, vol. 45, no. 4–6, pp. 117–229, 2002.

80. F. B. de Mongeot, M. Scherer, B. Gleich, E. Kopatzki, and R. J. Behm, "CO adsorption and oxidation on bimetallic Pt/Ru(0001) surfaces—a combined STM and TPD/TPR study," Surface Science, vol. 411, no. 3, pp. 249–262, 1998.

81. H. A. Gasteiger, N. Markovic, P. N. Ross, and E. J. Cairns, "Electro-oxidation of small organic molecules on well-characterized PtRu alloys," Electrochimica Acta, vol. 39, no. 11-12, pp. 1825–1832, 1994.

82. N. M. Markovic, H. A. Gasteiger, P. N. Ross, X. Jiang, I. Villegas, and M. J. Weaver, "Electro-oxidation mechanisms of methanol and formic acid on Pt-Ru alloy surfaces,"Electrochimica Acta, vol. 40, no. 1, pp. 91–98, 1995.

83. S. Desai and M. Neurock, "A first principles analysis of CO oxidation over Pt and Pt 66.7%Ru33.3% (111) surfaces," Electrochimica Acta, vol. 48, no. 25-26, pp. 3759–3773, 2003.

84. S. K. Desai and M. Neurock, "First-principles study of the role of solvent in the dissociation of water over a Pt-Ru alloy," Physical Review B, vol. 68, no. 7, Article ID 075420, pp. 754201–754207, 2003.

85.  B. E. Hayden, D. Pletcher, M. E. Rendall, and J. P. Suchsland, "CO oxidation on gold in acidic environments: particle size and substrate effects," Journal of Physical Chemistry C, vol. 111, no. 45, pp. 17044–17051, 2007.

86.  B. E. Hayden, D. Pletcher, and J. P. Suchsland, "Enhanced activity for electrocatalytic oxidation of carbon monoxide on titania-supported gold nanoparticles," Angewandte Chemie, vol. 46, no. 19, pp. 3530–3532, 2007.

87.  J. L. Roberts and D. T. Sawyer, "Electrochemical oxidation of carbon monoxide at gold electrodes," Electrochimica Acta, vol. 10, no. 10, pp. 989–1000, 1965.

88.  P. Rodriguez, N. Garcia-Araez, and M. T. M. Koper, "Self-promotion mechanism for CO electrooxidation on gold," Physical Chemistry Chemical Physics, vol. 12, no. 32, pp. 9373–9380, 2010.

89.  P. Rodriguez, N. Garcia-Araez, A. Koverga, S. Frank, and M. T. M. Koper, "CO electroxidation on gold in alkaline media: a combined electrochemical, spectroscopic, and DFT study," Langmuir, vol. 26, no. 14, pp. 12425–12432, 2010.

90.  I. X. Green, W. Tang, M. Neurock, and J. T. Yates Jr., "Spectroscopic observation of dual catalytic sites during oxidation of CO on a $Au/TiO_2$ catalyst," Science, vol. 333, no. 6043, pp. 736–739, 2011.

91.  M. Neurock, D. Hibbitts, and J. A. Dumesic, "Mechanistic insights into the role of water on the oxidation of CO over Au," to be submitted, 2011.

92.  N. Dimitratos, J. A. Lopez-Sanchez, J. M. Anthonykutty et al., "Oxidation of glycerol using gold-palladium alloy-supported nanocrystals," Physical Chemistry Chemical Physics, vol. 11, no. 25, pp. 4952–4961, 2009.

93.  N. Dimitratos, J. A. Lopez-Sanchez, S. Meenakshisundaram et al., "Selective formation of lactate by oxidation of 1,2-propanediol using gold palladium alloy supported nanocrystals," Green Chemistry, vol. 11, no. 8, pp. 1209–1216, 2009.

94.  M. D. Hughes, Y. J. Xu, P. Jenkins et al., "Tunable gold catalysts for selective hydrocarbon oxidation under mild conditions," Nature, vol. 437, no. 7062, pp. 1132–1135, 2005.

95. G. J. Hutchings, S. Carrettin, P. Landon et al., "New approaches to designing selective oxidation catalysts: Au/C a versatile catalyst," Topics in Catalysis, vol. 38, no. 4, pp. 223–230, 2006.

96. M. Sankar, N. Dimitratos, D. W. Knight et al., "Oxidation of glycerol to glycolate by using supported gold and palladium nanoparticles," ChemSusChem, vol. 2, no. 12, pp. 1145–1151, 2009.

97. B. N. Zope and R. J. Davis, "Influence of reactor configuration on the selective oxidation of glycerol over $Au/TiO_2$," Topics in Catalysis, vol. 52, no. 3, pp. 269–277, 2009.

98. B. N. Zope, D. D. Hibbitts, M. Neurock, and R. J. Davis, "Reactivity of the gold/water interface during selective oxidation catalysis," Science, vol. 330, no. 6000, pp. 74–78, 2010.

99. J. S. Spendelow, P. K. Babu, and A. Wieckowski, "Electrocatalytic oxidation of carbon monoxide and methanol on platinum surfaces decorated with ruthenium," Current Opinion in Solid State and Materials Science, vol. 9, no. 1-2, pp. 37–48, 2005.

100. J. S. Spendelow and A. Wieckowski, "Electrocatalysis of oxygen reduction and small alcohol oxidation in alkaline media," Physical Chemistry Chemical Physics, vol. 9, no. 21, pp. 2654–2675, 2007.

101. Z. Borkowska, A. Tymosiak-Zielinska, and G. Shul, "Electrooxidation of methanol on polycrystalline and single crystal gold electrodes," Electrochimica Acta, vol. 49, no. 8, pp. 1209–1220, 2004.

102. P. Parpot, A. P. Bettencourt, A. M. Carvalho, and E. M. Belgsir, "Biomass conversion: attempted electrooxidation of lignin for vanillin production," Journal of Applied Electrochemistry, vol. 30, no. 6, pp. 727–731, 2000.

103. P. Parpot, A. P. Bettencourt, G. Chamoulaud, K. B. Kokoh, and E. M. Belgsir, "Electrochemical investigations of the oxidation-reduction of furfural in aqueous medium—application to electrosynthesis," Electrochimica Acta, vol. 49, no. 3, pp. 397–403, 2004.

104. P. Parpot, N. Nunes, and A. P. Bettencourt, "Electrocatalytic oxidation of monosaccharides on gold electrode in alkaline medium: structure-reactivity relationship," Journal of Electroanalytical Chemistry, vol. 596, no. 1, pp. 65–73, 2006.

105. P. Parpot, S. G. Pires, and A. P. Bettencourt, "Electrocatalytic oxidation of D-galactose in alkaline medium," Journal of Electroanalytical Chemistry, vol. 566, no. 2, pp. 401–408, 2004.

106. P. Parpot, P. R. B. Santos, and A. P. Bettencourt, "Electro-oxidation of d-mannose on platinum, gold and nickel electrodes in aqueous medium," Journal of Electroanalytical Chemistry, vol. 610, no. 2, pp. 154–162, 2007.

107. S. H. Yan, S. C. Zhang, Y. Lin, and G. R. Liu, "Electrocatalytic performance of gold nanoparticles supported on activated carbon for methanol oxidation in alkaline solution," Journal of Physical Chemistry C, vol. 115, no. 14, pp. 6986–6993, 2011.

108. P. A. Christensen, A. Hamnett, and D. Linares-Moya, "Oxygen reduction and fuel oxidation in alkaline solution," Physical Chemistry Chemical Physics, vol. 13, no. 12, pp. 5206–5214, 2011.

109. D. Kardash, C. Korzeniewski, and N. Markovic, "Effects of thermal activation on the oxidation pathways of methanol at bulk Pt-Ru alloy electrodes," Journal of Electroanalytical Chemistry, vol. 500, no. 1-2, pp. 518–523, 2001.

110. A. V. Tripkovi, K. D. Popovi, B. N. Grgur, B. Blizanac, P. N. Ross, and N. M. Markovi, "Methanol electrooxidation on supported Pt and PtRu catalysts in acid and alkaline solutions," Electrochimica Acta, vol. 47, no. 22-23, pp. 3707–3714, 2002.

111. F. Colmati, G. Tremiliosi-Filho, E. R. Gonzalez, A. Berna, E. Herrero, and J. M. Feliu, "The role of the steps in the cleavage of the C-C bond during ethanol oxidation on platinum electrodes," Physical Chemistry Chemical Physics, vol. 11, no. 40, pp. 9114–9123, 2009.

112. F. Colmati, G. Tremiliosi-Filho, E. R. Gonzalez, A. Berna, E. Herrero, and J. M. Feliu, "Surface structure effects on the electrochemical oxidation of ethanol on platinum single crystal electrodes," Faraday Discussions, vol. 140, pp. 379–397, 2008.

113. V. Del Colle, A. Berna, G. Tremiliosi-Filho, E. Herrero, and J. M. Feliu, "Ethanol electrooxidation onto stepped surfaces modified by Ru deposition: electrochemical and spectroscopic studies," Physical Chemistry Chemical Physics, vol. 10, no. 25, pp. 3766–3773, 2008.

114. J. Souza-Garcia, E. Herrero, and J. M. Feliu, "Breaking the C-C bond in the ethanol oxidation reaction on platinum electrodes: effect of steps and ruthenium adatoms,"ChemPhysChem, vol. 11, no. 7, pp. 1391–1394, 2010.

115. S. C. S. Lai, S. E. F. Kleijn, F. T. Z. Ozturk et al., "Effects of electrolyte pH and composition on the ethanol electro-oxidation reaction," Catalysis Today, vol. 154, no. 1-2, pp. 92–104, 2010.

116. S. C. S. Lai, S. E. F. Kleyn, V. Rosca, and M. T. M. Koper, "Mechanism of the dissociation and electrooxidation of ethanol and acetaldehyde on platinum as studied by SERS," Journal of Physical Chemistry C, vol. 112, no. 48, pp. 19080–19087, 2008.

117. S. C. S. Lai and M. T. M. Koper, "Electro-oxidation of ethanol and acetaldehyde on platinum single-crystal electrodes," Faraday Discussions, vol. 140, pp. 399–416, 2008.

118. S. C. S. Lai and M. T. M. Koper, "Ethanol electro-oxidation on platinum in alkaline media," Physical Chemistry Chemical Physics, vol. 11, no. 44, pp. 10446–10456, 2009.

119. S. C. S. Lai and M. T. M. Koper, "The influence of surface structure on selectivity in the ethanol electro-oxidation reaction on platinum," Journal of Physical Chemistry Letters, vol. 1, no. 7, pp. 1122–1125, 2010.

120. M. Neurock, M. Janik, and A. Wieckowski, "A first principles comparison of the mechanism and site requirements for the electrocatalytic oxidation of methanol and formic acid over Pt," Faraday Discussions, vol. 140, no. 1, pp. 363–378, 2008.

121. S. R. Brankovic, J. X. Wang, Y. Zhu, R. Sabatini, J. McBreen, and R. R. Adzic, "Electrosorption and catalytic properties of bare and Pt modified single crystal and nanostructured Ru surfaces," Journal of Electroanalytical Chemistry, vol. 524-525, pp. 231–241, 2002.

122. T. Iwasita, "Electrocatalysis of methanol oxidation," Electrochimica Acta, vol. 47, no. 22-23, pp. 3663–3674, 2002.

123. W. C. Ketchie, M. Murayama, and R. J. Davis, "Selective oxidation of glycerol over carbon-supported AuPd catalysts," Journal of Catalysis, vol. 250, no. 2, pp. 264–273, 2007.

124. A. T. Governo, L. Proenca, P. Parpot, M. I. S. Lopes, and I. T. E. Fonseca, "Electro-oxidation of D-xylose on platinum and gold

electrodes in alkaline medium,"Electrochimica Acta, vol. 49, no. 9-10, pp. 1535–1545, 2004.

125. Y. Kwon and M. T. M. Koper, "Combining voltammetry with HPLC: application to electro-oxidation of glycerol," Analytical Chemistry, vol. 82, no. 13, pp. 5420–5424, 2010.

126. Z. W. Liu, F. Peng, H. J. Wang, H. Yu, W. X. Zheng, and J. A. Yang, "Phosphorus-doped graphite layers with high electrocatalytic activity for the $O_2$ reduction in an alkaline medium," Angewandte Chemie, vol. 50, no. 14, pp. 3257–3261, 2011.

127. A. C. Chen and J. Lipkowski, "Electrochemical and spectroscopic studies of hydroxide adsorption at the Au(111) electrode," Journal of Physical Chemistry B, vol. 103, no. 4, pp. 682–691, 1999.

128. S. Desai, "Theoretical investigation of solution effects on metal catalyzed hydrogenation and oxidation processes," in Department of Chemical Engineering, University of Virginia, 2002.

129. F. T. Wagner and T. E. Moylan, "Generation of surface hydronium from water and hydrogen coadsorbed on Pt(111)," Surface Science, vol. 206, no. 1-2, pp. 187–202, 1988.

130. N. Kizhakevariam and E. M. Stuve, "Coadsorption of water and hydrogen on Pt(100): formation of adsorbed hydronium ions," Surface Science, vol. 275, no. 3, pp. 223–236, 1992.

# Chapter 3

# Catalytic Technologies for Biodiesel Fuel Production and Utilization of Glycerol: A Review

Le Tu Thanh[1], Kenji Okitsu[2], Luu Van Boi[3], and Yasuaki Maeda[1]

[1]Research Organization for University–Community Collaborations, Osaka Prefecture University, 1-2 Gakuen-cho, Naka-ku, Sakai 599-8531, Japan

[2]Graduate School of Engineering, Osaka Prefecture University, 1-1 Gakuen-cho, Naka-ku, Sakai 599-8531, Japan

[3]Faculty of Chemistry, Vietnam National University, 19 Le Thanh Tong St., Hanoi, Vietnam

# ABSTRACT

More than 10 million tons of biodiesel fuel (BDF) have been produced in the world from the transesterification of vegetable oil with methanol by using acid catalysts (sulfuric acid, $H_2SO_4$), alkaline catalysts (sodium hydroxide, NaOH or potassium hydroxide, KOH), solid catalysts and enzymes. Unfortunately, the price of BDF is still more expensive than that of petro diesel fuel due to the lack of a suitable raw material oil. Here, we review the best selection of BDF production systems including raw materials, catalysts and production technologies. In addition, glycerol formed as a by-product needs to be converted to useful chemicals to reduce the amount of glycerol waste. With this in mind, we have also reviewed some recent studies on the utilization of glycerol.

# INTRODUCTION

After the disaster of Fukushima's nuclear power plant on 11th of March in 2011 in Japan, we should reconsider the role of atomic energy to protect global warming. Besides solar battery, wind power generation, and geothermal power generation, biomass energy resources such as methane, ethanol and BDF have attracted much attention as green energy for the mitigation of global warming due to the advantage of carbon neutrality of biomass. However, many scientists have been warning against the effectiveness of biomass energy. For example, with bio-ethanol produced in Brazil it has been pointed out that this is not mitigation but sometimes increases global warming because it is produced from plants cultivated at tropical forest area.

The term biofuel refers to solid (bio-char), liquid (ethanol and biodiesel), or gaseous (biogas, biohydrogen and biosynthetic gas) fuels that are predominantly produced from biomass. The most popular biofuels such as ethanol from sugar cane, corn, wheat or cassava and biodiesel from sunflower, soybean, canola are produced from food crops that require good quality land for plantation. However, ethanol can be produced from inexpensive cellulosic biomass resources such as herbaceous and woody plants from agriculture and forestry residues. Therefore, production of bioethanol from biomass is one excellent way to reduce raw material costs. In contrast, biodiesel production is the most popular one because the formation process is faster and the simpler compared with ethanol and methane production. There is also a growing interest in the use of waste cooking oil, and animal fats as cheap raw materials for biodiesel production [1,2].

Advantages of biofuels are the following: (a) biofuels are widely adapted with existing filling-fuel stations; (b) they can be used with current vehicles; (c) they are easily available from common biomass sources; (d) they are easily biodegradable; (e) they present a carbon-cycle in combustion; (f) there are many benefits to the environment, economy and consumers in using biofuels. Due to the reasons listed above, biofuels have become more attractive to several countries. Table 1 shows the main advantages of using biofuels [1,3].

**Table 1**: Major benefits of biofuels

| Environmental impacts | Reduction of green house gasses |
|---|---|
| | Reduction of air pollution |
| | Higher combustion efficiency |
| | Easily biodegradable |
| | Carbon neutral |
| **Energy security** | Domestically distributed |
| | Supply reliability |
| | Reducing use of fossil fuels |
| | Reducing the dependency on imported petroleum |
| | Renewable |
| | Fuel diversity |
| **Economic impacts** | Sustainability |
| | Increased number of rural manu-facturing jobs |
| | Increased farmer income |
| | Agricultural development |

Biofuels production has dramatically increased in the last two decades. Figure 1 shows the world production of ethanol and biodiesel between 2000 and 2010 [4]. In this stage, world ethanol production has increased from around 17 billion liters to 85 billion liters per year. Brazil was the world's leading ethanol producer until 2005 when USA roughly equaled Brazil but USA produced about twice that of Brazil in 2010. In contrast, Germany is the world's leader in biodiesel production with 30% of the world production. At present, since almost all liquid fuels are produced from food crops such as cereals, sugar cane and oil seeds, the raw materials supplied for biofuel production are limited. Therefore, to increase the yield of biofuels satisfying energy demand in the near future, it is necessary to find abundant inedible biomass such as agricultural residue, wood chip, industrial waste, *etc.* [5]. BDF has many advantages such as (1) high cetane number about 50; (2) built-in oxygen content; (3) burns fully; (4) no sulphur content; (5) no aromatics; (6) complete $CO_2$ cycle (carbon neutral in 1 year).

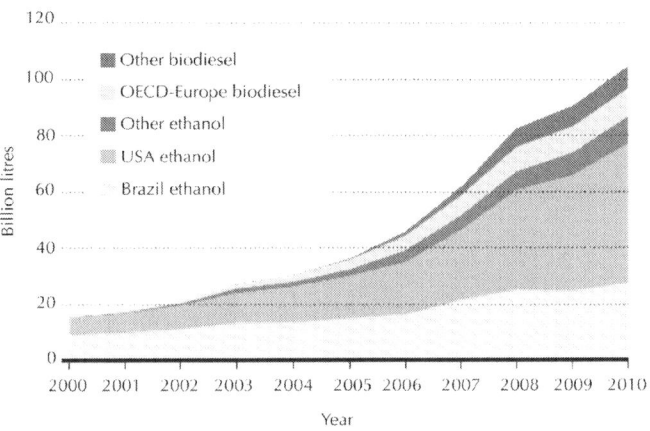

**Figure 1**: Global Biofuel Production. Reprinted with permission from [4]. Copyright OECD/IEA (2011).

BDF could be produced by adding methanol to waste cooking oil with small amounts of KOH or NaOH as a catalyst. However, some questions remain: (1) What is the best raw material available that does not increase food prices or deforestation? (2) What is the best production method for a green process by which fatty acid methyl ester (FAME) can be obtained with a minimal emission of waste and low energy consumption? One solution proposed to reduce the formation of soap with an alkaline catalyst was the application of an enzyme catalyst but the reaction rate was too slow. Another solution is the addition of solvent to the reaction mixture of oil and methanol to produce BDF in a homogeneous phase [6].

In general, there is no problem with alkaline catalyst processes with the use of good quality raw oil materials. If we use poor raw oil materials containing a high amount of free fatty acid (FFA) and moisture, we would need the excellent acidic catalyst of the esterification reaction of FFA and methanol. However, at present, the best catalyst might be still sulfuric acid at relatively high temperature. The most interesting scientific field of catalysts in biodiesel production is the transformation of glycerol to useful chemicals. In this review, we will briefly present the conventional catalysts and thriving technologies for the production of BDF as well as the new trends for utilization of the by-product glycerol.

# BIODIESEL PRODUCTION

## How to Produce Biodiesel?

The main components of vegetable oils and animal fats are triglycerides, which are esters of FFA with glycerol. The triglyceride typically contains several FFA, and thus different FFA can be attached to one glycerol backbone. With different FFA, triglyceride has different physical and chemical properties. The FFA composition is the most important factor influencing the corresponding properties of vegetable oils and animal fats. The fatty acid compositions of normal vegetable oils and fat are shown in Table 2, and the physical properties of oils, fat and petro-diesel are listed in Table 3 [6,7,8,9].

Because vegetable oils or animal fats have high viscosity, i.e., 35–50 mm$^2$ s$^{-1}$, it is necessary to reduce the viscosity in order to use them in a common diesel engine. There are four methods used to solve this problem: blending with petro-diesel, pyrolysis, microemusification (co-solvent blending) and transesterification. Among these methods, only the transesterification reaction creates the products commonly known as biodiesel [7].

Biodiesel can be synthesized by the transesterification reaction of a triglyceride with a primary alcohol in the presence of catalysts. Among primary alcohols, methanol is favored for the transesterification due to its high reactivity (the shortest alkyl chain and most polar alcohol) and the least expensive alcohol, except in some countries. In Brazil, for example, where ethanol is cheaper, ethyl esters are used as fuel. Furthermore, methanol has a low boiling point, thus excess methanol from the glycerol phase is easily recovered after phase separation [7].

The choice of a catalyst for the transesterification mainly depends on the amount of FFA and of raw materials. Table 4 shows the concentration of FFA in the representative oils. If the oils have high FFA content and water, the acid-catalyst transesterification process is preferable. However, this process requires relatively high temperatures, *i.e.*, 60–100 °C, and long reaction times, *i.e.*, 2–10 h, in addition to causing undesired corrosion of the equipment. Therefore, to reduce the reaction time, the process with an acid-catalyst is adapted as a pre-treatment step only when necessary to convert FFA to esters. Then, the addition of an alkaline-catalyst is followed for the transesterification step to transform triglycerides to esters [10,11]. In contrast, when the FFA content in the oils is less than one wt.%, many researchers have recommended that only an alkaline-catalyst assisted process should be applied, because this process requires less and simpler equipment than that for the case of higher FFA content mentioned above.

**Table 2:** Major fatty acids in oils and fat [6,7,8,9]

| Oils and fat | Iodine value | Soponification value | Fatty acid composition (wt.%) | | | | | | | | |
|---|---|---|---|---|---|---|---|---|---|---|---|
| | | | 10:0 | 12:0 | 14:0 | 16:0 | 18:0 | 18:1 | 18:2 | 18:3 | 22:1 |
| *Oils* | | | | | | | | | | | |
| Canola | 109–126 | 188–193 | - | - | - | 2.5–5.7 | 1.15–2.4 | 52–61.9 | 15.1–22.3 | 6.4–11.7 | 0.8–1.6 |
| Olive | 75–94 | 184–196 | - | 0–1.3 | 7–20 | 0.5–5 | 55–84.5 | 3.5–21 | - | - | - |
| Corn | 103–140 | 187–198 | - | - | 0–0.3 | 7–16.5 | 1–3.3 | 20–43 | 39–62.5 | 0.5–1.5 | - |
| Catfish | 31–57 | 187–192 | - | - | 2.0–3.5 | 21.2–27.4 | 7.1–9.3 | 45.1–48.0 | 12.0–16.0 | 1.0–2.3 | 0.3–0.5 |
| Cottonseed | 9–119 | 189–198 | - | - | 0.6–1.5 | 21.4–26.4 | 2.1–5 | 14.7–21.7 | 46.7–58.2 | | - |
| *Jatropha curcas* | 92–112 | 177–189 | - | - | 0.3–0.4 | 12.6–14.2 | 5.97–6.9 | 39.5–44.1 | 34.4–37.8 | 2.4–3.4 | 0.5–0.7 |
| Palm | 35–61 | 186–209 | 0–0.4 | 0.5–2.4 | 32–47.5 | 36–53 | 3.5–6.3 | 6–12 | - | - | - |
| Peanut | 80–106 | 187–196 | - | - | 0–0.5 | 6–14 | 1.9–6 | 36.4–67.1 | 13–43 | - | 0–0.3 |
| Rapeseed | 94–120 | 168–187 | - | - | 0–1.5 | 1–6 | 0.5–3.5 | 8–60 | 9.5–23 | 1–13 | 5–64 |
| Soybean | 117–143 | 189–195 | - | - | - | 4.3–13.3 | 2.4–6 | 17.7–30.8 | 49–57.1 | 2–10.5 | 0–0.3 |
| Sunflower | 110–143 | 186–194 | - | - | - | 3.5–7.6 | 1.3–6.5 | 14–43 | 44–74 | - | - |
| *Fat* | | | | | | | | | | | |
| Tallow | 35–48 | 218–235 | - | - | 2.1–6.9 | 25–37 | 9.5–34.2 | 14–50 | 26–50 | - | - |

Note: ᵃ(Carbon number: Double bond)

**Table 3**: Physical properties of oils, fat and petro-diesel [7,8]

| Oils, fat and petro-diesel | Cetane number | Kinematic viscosity (37.8 °C, mm$^2$ s$^{-1}$) | Flash point (°C) |
|---|---|---|---|
| *Oils* | | | |
| Corn | 37.6 | 34.9 | 277 |
| Cottonseed | 41.8 | 33.5 | 234 |
| *Jatropha curcas* | 38.0 | 37.0 | 240 |
| Peanut | 41.8 | 39.6 | 271 |
| Rapeseed | 37.6 | 37.0 | 246 |
| Soybean | 37.9 | 32.6 | 254 |
| Sunflower | 37.1 | 37.1 | 274 |
| *Fat* | | | |
| Tallow | - | 51.2 | 201 |
| *Petro-diesel* | | | |
| Diesel fuel No. 2 | 47.0 | 2.7 | 52 |

**Table 4**: Acid value in representative oils

| Oils and Fats | Acid value mg KOH/1 g oil | References |
|---|---|---|
| Refined sunflower | 0.2–0.5 | [12,13] |
| Crude *Jatropha curcas* | 15.6–43 | [8,14] |
| Refined Safflower | 0.35 | [15] |
| Crude palm | 6.9–50.8 | [16,17] |
| Cottonseed | 0.6–2.87 | [18,19] |
| Corn | 0.1–5.72 | [20,21] |
| Coconut | 1.99–12.8 | [22,23] |
| Soybean | 0.1–0.2 | [24,25] |
| Animal fats | 4.9–13.5 | [26] |
| Canola | 0.6–0.8 | [27,28] |
| Waste cooking | 0.67–3.64 | [29] |

Several reviews dealing with the production of biodiesel by transesterification have been published [10,30]. Commonly, the transesterification can be catalyzed by a base or acid-catalyst. The triglyceride is converted stepwise to diglyceride and monoglyceride intermediates, and finally to glycerol [31]. Mechanisms of the transesterification of triglyceride with alcohol in the presence of a base or acid-catalyst are shown as follows:

## Base-catalyst *[32]:*

$$ROH + B \rightleftharpoons RO^{\ominus} + BH^{\oplus} \qquad (1)$$

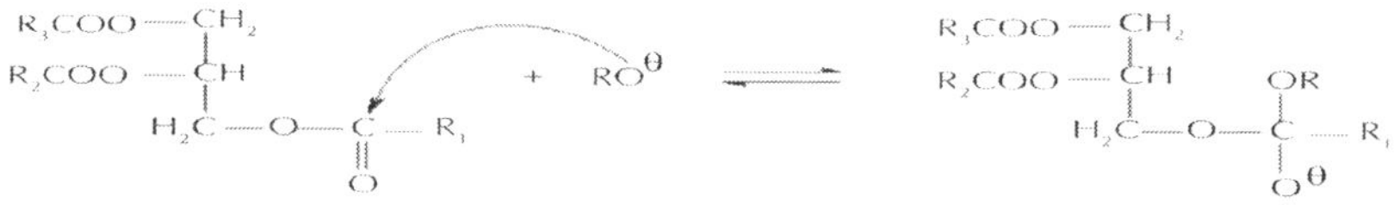

$$(2)$$

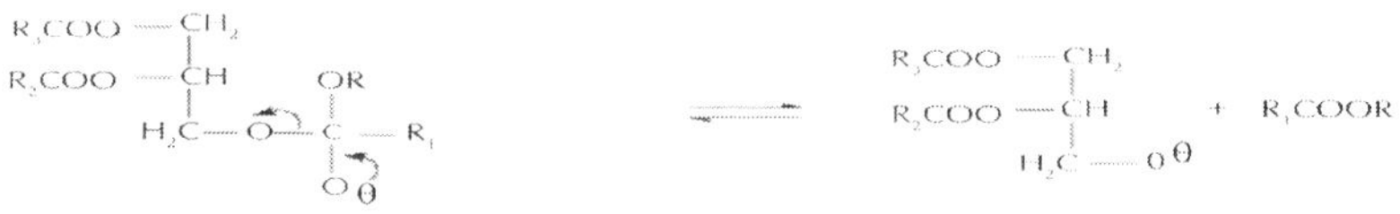

$$(3)$$

$$(4)$$

## Acid-catalyst[33]:

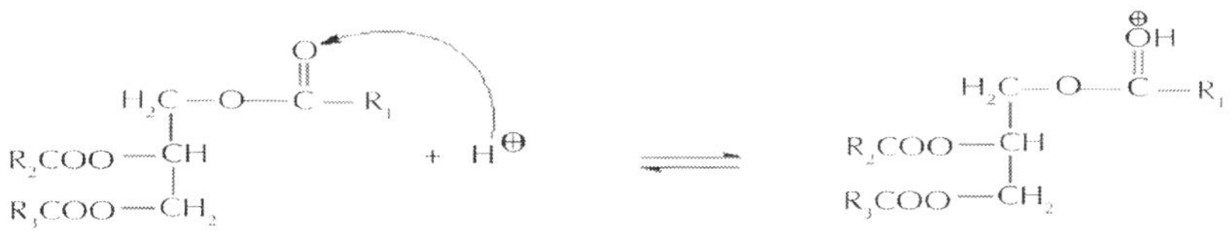

$$(5)$$

$$(6)$$

$$(7)$$

$$(8)$$

These reactions demonstrate the conversion of triglyceride into diglyceride. The reaction mechanisms of diglyceride and monoglyceride, which convert into monoglyceride and glycerol, respectively, take place in the same way as for triglyceride. The overall reactions are shown as follows:

$$(9)$$

where $R$, $R_1$, $R_2$ and $R_3$ are alkyl groups.

## Possible Methods for Biodiesel Production

It is believed that the transesterification process includes three stages: (1) the mass transfer between oil and alcohol; (2) the transesterification reaction; and (3) the establishment of equilibrium. Because alcohol and oil are immiscible, mixing efficiency is one of the most important factors to improve the yield of transesterification. Therefore, this section focuses on methods that can improve the efficiency of the mass transfer between the reactants. There are many adaptable methods to conduct transesterification such as mechanical stirring, supercritical alcohol, ultrasonic irradiation, *etc.* [34,35,36,37,38,39]. More details of each method will be demonstrated in the followings sections.

## *Mechanical Stirring Method*

Normally, the transesterification of a triglyceride with alcohol in the presence of a catalyst is carried out in a batch reactor. At first, the reactants are heated up to a desired temperature, and then they are mixed well by a mechanical stirring tool. The fatty acid methyl ester (FAME) yield is dependent on various parameters such as type and amount of the catalyst, reaction temperature, ratio of alcohol to oil, mixing intensity, *etc.* The mechanical stirring method, a popular one for BDF production, is suitable for both homogeneous and heterogeneous catalysts. This method is described as follows.

## *Homogeneous Base-Catalyst Transesterification*

The transesterification reaction is catalyzed by alkaline metal hydroxides or alkoxides, as well as sodium or potassium carbonates. The alkaline catalysts give good performance when raw materials with high quality (FFA < 1 wt.% and moisture < 0.5 wt.%) are used [40]. The reaction is carried out at a temperature of 60–65 °C under atmospheric pressure with an excess amount of alcohol, usually methanol. The molar ratio of alcohol to oil is often 6:1 or more. This ratio is two-times higher or more than the stoichiometric ratio of alcohol given in the reaction scheme (9) as described above. It often takes several hours to complete the reaction when alkaline hydroxides such as NaOH or KOH are used. Alkaline alkoxides, e.g., sodium alkoxide, are the most reactive catalysts because the yield of FAME that can be attained is higher than 98% in a short reaction time of 30 min. Alkaline hydroxides are cheaper than the alkaline alkoxides, but less active. The yield of FAME can be improved by simply increasing the amount of the alkaline hydroxides by one or two mol% to oil, and thus they are a good alternative to the alkaline alkoxides [41]. Sivakumar *et al.* produced BDF from raw material dairy waste scum and the FAME yield reached 96.7% under the optimal conditions: KOH 1.2 wt.%; molar ratio of methanol to oil 6:1; reaction temperature 75 °C; reaction time 30 min at 350 rpm [42].

One of the biggest drawbacks for the base-catalyst is that it cannot be applied directly when the oils or fats contain large amounts of FFA, *i.e.*, >1 wt.%. Since the FFA is neutralized by the base catalyst to produce soap and water, the activity of the catalyst is decreased. Additionally, the formation of soap inhibits the separation of glycerol from the reaction mixture and the purification of FAME with water [43]. Removal of these saponified catalysts is technically difficult and it adds extra cost to the production of biodiesel. Furthermore, since homogeneous base catalysts mainly dissolve in the glycerol and alcohol phase after the reaction is completed, they cannot be recycled for the following batches, and the crude BDF must be purified by a washing process with water or a distillation at high temperature under reduced pressure.

In consequence, with vegetable oils or fats containing low FFA and water, the base-catalyst transesterification is much faster than the acid-catalyst transesterification and is most commonly used commercially on the industrial scale [44].

## Homogeneous Acid-Catalyst Transesterification

With starting raw materials containing a high amount of FFA such as waste cooking, *Jatropha curcas*, rubber, tobacco oils, *etc.*, an acid-catalyst, usually a strong acid such as sulfuric, hydrochloric or phosphoric acid, is more favorable than base-catalyst because the reaction does not form soap. However, the acid-catalyst is very sensitive to the water content of the raw materials. It was reported that a small amount of water, *i.e.*, 0.1 wt.% in the reaction mixture affected the FAME yield of the transesterification of vegetable oil with methanol. If the concentration of water is 5 wt.%, the reaction is completely inhibited. Canakci and Gerpen conducted simultaneous esterification and transesterification reactions with acid catalysts where the yield of FAME attained was more than 90% with water content of less than 0.5 wt.% under the reaction conditions of temperature 60 °C; molar ratio of methanol to oil 6:1; sulfuric acid 3.0 wt.%, and reaction time 96 h [45].

Disadvantages of the acid-catalyst are that they require higher temperature and longer reaction time, in addition to causing undesired corrosion of the equipment. Moreover, to increase the conversion of triglyceride, a large excess amount of methanol, e.g., molar ratio of methanol to oil of higher than 12:1, should be used. In practice, therefore, to reduce the reaction time, the process with an acid-catalyst is adapted as a pretreatment step only when it is necessary to convert FFA to esters, and is followed by a base-catalyst addition for the transesterification step to transform triglyceride to esters. In general, acid-catalyst transesterification is usually performed at the following conditions: a high molar ratio of methanol to oil of 12:1; high temperatures of 80–100 °C; and a strong acid namely sulfuric acid [10]. Patil et al. performed a two-step process for production of BDF from *Jatropha curcas* oil with a maximum yield of 95% attained according to the reaction conditions: at the first acid esterification, *i.e.*, methanol to oil molar ratio of 6:1, sulfuric acid of 0.5 wt.%, and reaction temperature of 40 ± 5 °C; followed by alkaline transesterification with methanol to oil molar ratio of 9:1, KOH of 2 wt.%, and reaction temperature of 60 °C [46].

## Heterogeneous Solid-Catalyst Transesterification

As mentioned above, the disadvantages of homogeneous base-catalyst transesterification are high energy-consumption, costly separation of the catalyst from the reaction mixture and the purification of crude BDF. Therefore, to reduce the cost of the purification process, heterogeneous solid catalysts such as metal oxides, zeolites, hydrotalcites, and -alumina, have been used recently, because these catalysts can be easily separated from the reaction mixture, and can be reused. Most of these catalysts are alkali or alkaline oxides supported on materials with a large surface area. Similar to homogeneous catalyst, solid base-catalysts are more active than solid acid-catalysts [47,48]. In this review, we focus on popular solid base and acid catalysts.

## Activated Oxides of Calcium and Magnesium

Oxides of alkaline earth metals such as Be, Mg, Ca, Sr and Ba have been used for synthesis of BDF in several studies. CaO and MgO are abundant in nature and widely used among alkaline earth metals [49,50,51,52,53]. Ngamcharussrivichai *et al.* calcined domomite, mainly consisting of $CaCO_3$ and $MgCO_3$, at 800 °C for 2 h to prepare CaO and MgO catalysts for the transesterification of palm kernel oil. Under the optimal reaction conditions: amount of catalyst of 6 wt.% based on oil; molar ratio of methanol to oil of 30:1; reaction time of 3 h and reaction temperature of 60 °C, the yield of FAME was 98%. After each run, the catalyst was recovered by centrifuge and washed with methanol, and used for the next run. The results showed that the yield of FAME was more than 90% up to the seventh repetition [54]. Huaping *et al.* carried out the transesterification of *Jatropha curcas* oil with methanol catalyzed by calcium oxide, and the yield of FAME was higher than 93% under the conditions namely the catalyst amount of 1.5 wt.%; temperature of 70 °C; molar ratio of 9:1; and reaction time 3.5 h [55]. The activity of the solid catalyst is dependent on the active sites on the surface of CaO or MgO. Since the surface of these metal oxides is easily poisoned by absorption of carbon dioxide and water in the air to form carbonates and hydroxides, respectively, the activity of these catalysts decreases with time. However, the catalytic activity of these metal oxides can be recovered by calcination of the catalysts to remove carbon dioxide and water at high temperature. Grandos *et al.* activated CaO, which was exposed to the air for 120 days, at temperatures of 473 K, 773 K and 973 K, respectively. Figure 2 shows the yield of FAME with the CaO catalyst activated at different temperatures. The CaO catalyst pretreated by evacuation at 473 K gave a very low activity. The evacuation of the catalyst at 773 K can improve the catalytic activity due to dehydration of the $Ca(OH)_2$ present in the CaO catalyst. The best catalytic activation can be attained at 973 K due to the transformation of the $CaCO_3$ to CaO [56].

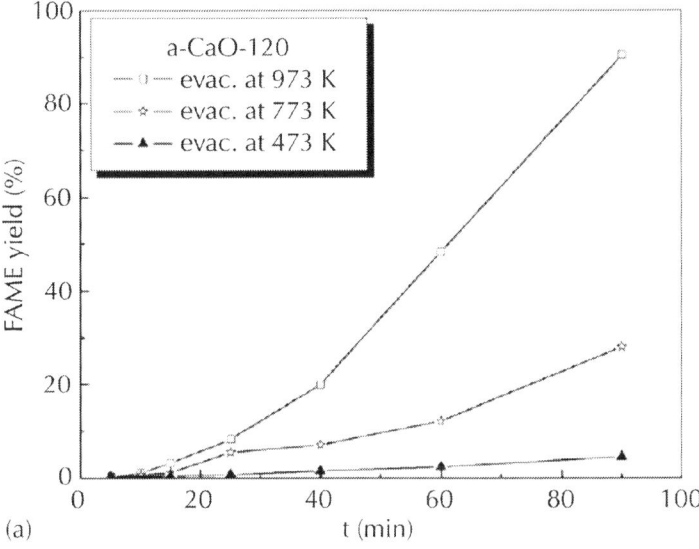

(a)

**Figure 2**: Effect of activated temperature and time of CaO catalyst on the fatty acid methyl ester (FAME) yields (Notes: a-CaO-120 means that the fresh CaO was exposed to room air for 120 days; evac. at 473 K, activated at 473 K). The reaction conditions: sunflower oil; catalyst amount to oil, 1 wt.%; molar ratio of methanol to oil, 13:1, temperature, 333 K; reaction time 100 min at 1000 rpm. Reprinted with permission from [56]. Copyright 2007 Elsevier.

## Alkaline Modified Zirconia Catalyst

Omar *et al.* studied alkaline modified zirconia catalysts such as Mg/ZrO$_2$, Ca/ZrO$_2$, Sr/ZrO$_2$, and Ba/ZrO$_2$ as heterogeneous catalysts for biodiesel production from waste cooking oil. The catalysts were prepared via wet impregnation of alkaline nitrate salts supported on zirconia. Among the tested catalysts, Sr/ZrO$_2$ had the highest catalytic activity. The active sites of the Sr/ZrO$_2$ can assist simultaneous esterification and transesterification reactions in the ethanolysis process. About 79.7% ME yield can be attained at 2.7 wt.% catalyst loading (Sr/ZrO$_2$), 29:1 of methanol ratio to oil, for 169 min and at 115.5 °C which was determined as the optimal reaction conditions [57].

## Tri-Potassium Phosphate

The transesterification of waste cooking oil with methanol, using solid catalysts such as tri-potassium phosphate ($K_3PO_4$), KOH and tri-sodium phosphate ($Na_3PO_4$), was investigated by Guan *et al.* Among the tested catalysts, $K_3PO_4$ showed the highest catalytic activity for the transesterification reaction. $K_3PO_4$ was hydrolyzed in the presence of water, and $HPO_4^{2-}$, $H_2PO_4^-$ and $OH^-$ ions were formed in the reaction solution. As a result, the reaction mixture showed a strong alkaline property. The FAME yield reached 97.3% when the transesterification was performed with a catalyst concentration of 4 wt.% at 60 °C for 120 min. The used $K_3PO_4$ was regenerated using an aqueous KOH solution. A FAME yield of 88% could be achieved when the regenerated catalyst was used [58].

## Metal Oxides Supported on Silica

Jacobson *et al.* synthesized and utilized various solid acid catalysts such as $MoO_3/SiO_2$, $MoO_3/ZrO_2$, $WO_3/SiO_2$, $WO_3/SiO_2–Al_2O_3$, zinc stearate supported on silica, zinc ethanoate supported on silica and 12-tungstophosphoric acid (TPA) supported on zirconia. They were synthesized and evaluated for biodiesel preparation from waste cooking oil containing 15 wt.% FFA. The results revealed that the zinc stearate immobilized on silica gel (ZS/Si) was the most effective catalyst in simultaneously catalyzing the transesterification of triglycerides and esterification of FFA present in waste cooking oil to methyl esters. The maximum FAME yield of 98 wt.% was obtained at the optimal parameters: molar ratio of methanol to oil of 18:1; catalyst amount of 3 wt.%; stirring speed of 600 rpm and reaction temperature of 200 °C with the most active ZS/Si catalyst. Particularly, the catalyst was recycled and reused many times without any loss in activity [59].

## Mixed Oxides of $TiO_2$-MgO

Wen *et al.* used mixed oxides of $TiO_2–MgO$ produced by the sol–gel method to convert waste cooking oil into biodiesel. The best catalyst was MT-1-923 comprising a Mg/Ti molar ratio of 1 and calcined at 650 °C. The main reaction parameters such as methanol/oil molar ratio, catalyst amount, and temperature were investigated. The best yield of FAME 92.3% was obtained at a molar ratio of methanol to oil of 50:1; catalyst amount of 10 wt.%; reaction time of 6 h and reaction temperature of 160 °C. They observed that the catalytic activity of MT-1-923 decreased slowly in the recycle process. To improve catalytic activity, MT-1-923 was regenerated by a two-step washing method (the catalyst was washed with methanol four times and subsequently with n-hexane once before being dried at 120 °C). The FAME yield slightly increased to 93.8% compared with 92.8% for the fresh catalyst due to an increase in the specific surface area and average pore diameter. The mixed oxides catalyst, $TiO_2$–MgO, showed good potential in large-scale biodiesel production from waste cooking oil [60].

## Solid Acid-Catalysts

Despite lower activity, solid acid catalysts have been used in many industrial processes because they contain a variety of acid sites on their surfaces with different strengths of Brönsted or Lewis acidity, compared to the homogenous acid-catalysts. Solid acid-catalysts such as Nafion-NR50, sulfated zirconia and tungstated zirconia were chosen to catalyze biodiesel-forming transesterification due to the presence of sufficient acid site strength [61]. Sulfonic acid ion-exchange resins have been reported to show excellent catalytic activity in esterification reaction as a pretreatment step for oils containing a high amount of FFA [62,63]. In a pioneering study, Santacesaria et al. studied the kinetics of esterification of a mixture of triglyceride and oleic acid (with initial acidity in the range of 47.1–58.3 wt.%) with methanol using an acid ion-exchange polymeric resin (2 wt.%) as the heterogeneous catalyst. The sulfonic acid resin displays an active catalyst for esterification with the conversion of oleic acid to methyl oleate reaching more than 80% within 2 h reaction time at 85 °C [64]. Melero et al. performed the transesterification of refined and crude vegetable oils with a sulfonic acid-modified mesostructured catalyst resulting in a yield of FAME purity of over 95 wt.%, for oil conversion close to 100%, under the best reaction conditions: temperature 180 °C, methanol/oil molar ratio 10, and catalyst loading 6 wt.% with regard to the amount of oil. They found that these sulfonated mesostructured materials are promising catalysts for the preparation of biodiesel; however, some aspects related to the adsorption properties of the silica surface and the enhancement of the catalyst's reusability need to be addressed [65].

Recently, promising catalysts based on biomass pyrolysis by-products (sugars, biochar, flyash, *etc.*) have been developed for production of biodiesel [66,67,68,69,70]. Hara *et al.* sulphonated incompletely carbonized natural products such as sugars, starch or cellulose resulting in a rigid carbon material. They used the solid sulphonated carbon catalyst to produce high-grade biodiesel. The results revealed that the activity of their catalyst is more than half again compared with that of a liquid sulfuric acid catalyst and much higher than that of conventional solid acid catalyst, and there was no loss of activity or leaching of $-SO_3H$ group during the process. In addition to this, the use of biomass materials is inexpensive and ecologically friendly [66]. Zong *et al.* successfully conducted the esterification of FFAs such as oleic, palmitic and stearic acids with methanol with a D-glucose-derived catalyst. The yields of FAME were higher than 95% under the reaction conditions: 10 mmol FFA; 100 mmol methanol; 0.14 g sugar catalyst; reaction temperature 80 °C [69].

## Enzyme-Catalyst Transesterification

The use of lipases as enzyme-catalysts for biodiesel production is also increasingly interesting [71]. The main purpose is to overcome the issues involving recovery and treatment of the by-products that requires complex processing apparatus [72]. The main drawback of the enzyme-catalyzed process is the high cost of the lipases. In order to reduce the cost, enzyme immobilization has been studied for ease of recovery and reuse [73]. Additionally, inactivation of the enzyme that leads to decrease of yields is mostly restricted by the low solubility of the enzyme in methanol [74]. Although lipase catalyzed transesterification offers an attractive alternative, the industrial application of this technology has been slow due to feasibility aspects and some technical challenges [40].

For instance, the optimized reaction conditions for the transesterification of tallow were as follows: temperatureof 45 °C; stirring speed of 200 rpm; enzyme concentrations of 12.5–25%, based on triglyceride; molar ratio of methanol to oil of 3:1, and reaction time 4–8 h (for primary alcohols) and 16 h (for secondary alcohols). Lipozyme, *i.e.*, IM 60 was most effective for the transesterification of tallow with a conversion of 95% when primary alcohols were used. In contrast, lipase from *C.antarctica* and *P. cepacia*(PS-30) was the most efficient with a conversion of 90% when secondary alcohols were used [75].

# Ultrasonic Irradiation Method

Since chemical and physical properties of vegetable oils are quite different from methanol, they are completely immiscible. The mass transfer between these reactants is one of the most important parameters affecting the yield of FAME. Ultrasonic irradiation is known to be a useful tool for strengthening mass transfer in liquid-liquid heterogeneous systems [36]. With increased liquid-liquid mass transfer, oils and methanol are easily mixed together. When sound waves with a suitable frequency are transmitted effectively from a transducer to liquids of oil and alcohol, a number of cavitation bubbles are formed in the liquids. The formation and collapse of cavitation bubbles disrupt the phase boundary in a two-phase liquid system. Owing to this aspect, alcohol and oil form easily a fine emulsion, where the droplet size of methanol and oil is in micrometers. As a result, the interface area of droplets of alcohol and oil is increased, and thus the transesterification reaction proceeds effectively. Under ultrasonic irradiation, therefore, the transesterification can be carried out at lower temperature with smaller amounts of catalyst and methanol compared with the conventional mechanical stirring method.

Since a low frequency of ultrasound gives a high mixing efficiency, the frequency adapted for biodiesel production is in the range from 20 to 40 kHz. Many researchers have studied the production of biodiesel in a laboratory scale using an ultrasonic water bath with frequency of 24, 28 and 40 kHz [76,77,78,79,80].

There are several types of transducers used for biodiesel production such as ultrasonic horn transducers, push-pull ultrasonic transducers, multiple transducers equipped to a water bath, *etc.* [81,82]. The ultrasonic-assisted transesterification can be carried out in batch or continuous reactors. Batch reactors using water bath or small horn type transducers are suitable for small capacities with a reactor volume in the range of 0.1–1 L [83,84,85,86]. Therefore, the batch transesterification process cannot be applied for production of biodiesel on large industrial scales. On the other hand, the reactor for the continuous process usually uses the horn type high power transducer with a capacity of 1–3 kW, and the transducer is connected to a reactor with volume of 1–3 L. Oil, methanol and catalyst are continuously introduced to the reactor by a pump system. Furthermore, the continuous separation and purification processes can be operated automatically when a continuous reactor is used [9,11]. Therefore, the continuous reactor is favorable for the production of biodiesel on a large industrial scale.

Since the ultrasonic irradiation method gives strong mixing effects, the reaction can be carried out at ambient temperature. Therefore, it is supposed that acid or base homogeneous catalysts are both suitable for the esterification and transesterification reaction [36,76]. Hanh et al. reported the esterification of oleic acid with several alcohols (ethanol, propanol and butanol) in the presence of $H_2SO_4$ in a batch reactor at temperatures of 10–60 °C, molar ratios of alcohol to oleic acid of 1:1–10:1, amount of catalysts of 0.5–10% based on oleic acid weight and irradiation times of 0.5–10 h. The optimum conditions for the esterification process were molar ratio of alcohol to oleic acid of 3:1; 5 wt.% of $H_2SO_4$ at 60 °C and irradiation time of 2 h [83]. Recently, Mootabadi et al. performed the transesterification of palm oil with methanol in the presence of alkaline earth metal oxide catalysts (CaO, BaO and SrO) in a batch process assisted by 20 kHz ultrasonic irradiation. They revealed that catalytic activity was in the sequence of CaO < SrO < BaO. The yields achieved in 10–60 min reaction times increased from 5.5% to 77.3% (CaO), 48.2% to 95.2% (SrO), and 67.3% to 95.2 (BaO) under the following reaction conditions: molar ratio of methanol to oil of 9:1; catalyst amount of 3 wt.%; and reaction temperature 65 °C [85].

Georgogianni et al. carried out the transesterification from waste oil in the presence of alkaline catalysts and that from soybean frying oils in the presence of other heterogeneous catalysts, using ultrasonic irradiation of 24 kHz and mechanical stirring of 600 rpm. Their results showed many advantages of ultrasonic irradiation such as high yield of FAME, time saving procedure, etc. compared to the mechanical stirring method [2,34]. Other studies on the transesterification of various vegetable oils with different types of alcohols in the presence of a base-catalyst have been published. Maeda et al. reported that the yield of FAME was greater than 95% within a 20 min reaction time at room temperature on the laboratory scale [82,86].

In order to apply the ultrasonic technique for larger scale production, Thanh et al. designed a pilot plant using the horn type transducer with a capacity of 1 kW and frequency of 20 kHz for production of biodiesel from canola oil and methanol. This system was carried out by a circulation process with a tank volume of 100 L. The high yield of FAME obtained was more than 99% under the following optimal conditions: molar ratio to oil 5:1, and KOH catalyst 0.7 wt.%, reaction time 1 h at ambient temperature. However, it was quite difficult to scale up this system to hundreds or thousands of liters because the methanol and glycerol separate from the reaction mixture and make the mixture non-uniform in the circulation tank [9]. Then, Thanh et al. attempted to modify the circulation reaction system to a continuous reaction system in order to adapt for large scale production. The experimental setup for the transesterification and purification is schematically depicted in Figure 3 [11]. The transesterification of waste cooking oil with methanol in the presence of KOH catalyst was carried out in the continuous ultrasonic reactor by a two-step process. The effects of the residence time of reactants in the reactor, molar ratio of methanol to waste cooking oil and separation time of glycerol from the reaction mixture in each step were investigated. It was found that the optimal conditions for the transesterification were the total molar ratio of methanol to oil 4:1, KOH 1.0 wt.%, and a residence time in the reactor of 56 s for the entire process. Under these conditions, the recovery of biodiesel from waste cooking oil is 93.83 wt.%. The properties of the product satisfy the Japanese Industrial Standard for biodiesel B100 (JIS K2390). This process significantly reduces the use of methanol compared to conventional methods (the mechanical stirring and supercritical methanol methods), which need a molar ratio of methanol to oil of at least 6:1. Therefore, the continuous ultrasonic reactor with a two-step process would be a beneficial technique for the production of biodiesel from waste cooking oil.

## Supercritical Alcohol Method

As a catalyst free method for transesterification uses, a supercritical methanol method has been investigated at high pressure (around 80 atm) and high temperatures (300–400 °C) in a continuous reactor. Under the supercritical condition, the reaction mixture becomes a single phase, and the reaction takes place rapidly and spontaneously [87]. Compared to processes using catalysts, the supercritical method has three main advantages as follows:

The first, this process is friendly for the environment, because no catalyst is needed in the reaction, therefore, the separation process of the catalyst and soap from alkyl esters is unnecessary. The second, the supercritical reaction has a shorter reaction time, *i.e.*, 2–4 min, than conventional methods using catalysts, and the conversion rate is very high [88]. The third, neither FFA nor the water content influences the reaction in the supercritical method. The FFA is converted to FAME instead of soap. Therefore, this process can be applied to a wide variety of feedstocks [89]. However, the disadvantages of the supercritical methods stem mainly from the high pressure and temperature requirement, and the high molar ratio of methanol to oil (usually 42:1) that makes the cost of the production process expensive [5].

To conduct the transesterification in the supercritical condition under a lower temperature, Demirbas carried out the reaction of sunflower oil with methanol in the presence of CaO catalyst in supercritical methanol for biodiesel production. The results revealed that the transesterification was essentially completed within 6 min with an amount of CaO catalyst of 3 wt.%, molar ratio methanol to oil 41:1 at 525 K instead of a temperature of more than 600 K in the case without catalyst [49].

## Co-Solvent Method

In order to conduct the reaction in a single phase, co-solvents such as tetrahydrofuran (THF), 1,4-dioxane and diethyl ether were examined. Among co-solvents listed above, THF was the first solvent used for the transesterification. At a molar ratio of methanol to oil of 6:1, the addition of THF 1.25 volumes to methanol into oil produced a one phase system in which the transesterification process was speeded up dramatically. Moreover, THF is chosen because its boiling point (67 °C) is only two degrees higher than that of methanol. Therefore, the excess methanol and THF can be co-distilled and recycled [6].

The transesterification of soybean oil with methanol was carried out at different concentrations of NaOH catalyst using co-solvent THF. The FAME yields were 82.5, 85, 87 and 96% obtained at catalyst concentrations of 1.1, 1.3, 1.4 and 2.0 wt.%, respectively, for a reaction time of 1 min. Similarly, for the transesterification of coconut oil using THF/methanol volume ratio 0.87 with NaOH of 1 wt.%, the conversion was 99% in 1 min [37].

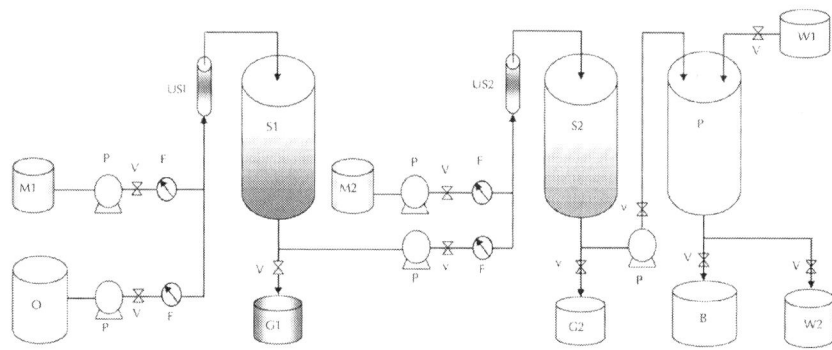

**Figure 3**: Flow diagram of an ultrasound-assisted continuous reactor for biodiesel production through a two-step process on the pilot plant. O: Oil tank; M1, M2: Methanol and catalyst tanks; P: Liquid pumps; V: Valve; F: Flow meters; US1, US2: Ultrasonic reactors; S1, S2: Separation tanks; G1, G2: glycerol tanks; P': Purification tank; B: Biodiesel production tank; W1, W2: fresh and waste water tanks. Reprinted with permission from [11]. Copyright 2010 Elsevier.

Recently, Maeda *et al.* presented the transesterification of vegetable oils and methanol in the presence of KOH catalyst by using several solvents such as acetone, THF, acetonitrile, diethyl ether, iso-propanol, *etc.* The transesterification assisted by the solvents shows the following new results: (1) the formation of FAME is completed even with smaller amounts of methanol added to oil (4 moles methanol to 1 mole oil), KOH catalyst (0.1–0.5 wt.% to oil) at room temperature; (2) the formation of soap is negligibly small due to the small amount of catalyst used and the reaction at ambient temperature; and (3) the separation rate of FAME with the by-product glycerol is speeded up more than 10 times compared with the conventional mechanical stirring method. In the case of acetone, which does not dissolve glycerol, the separation of FAME from glycerol was very fast because of the lower viscosity of the FAME-acetone solution and the larger difference between the low-density FAME-acetone solution.

Surprisingly, the formation of FAME was not retarded in the co-solvent method even in the presence of 5 wt.% of water as shown in Figure 4. In contrast, the yield of FAME at 60 min became *ca.* 15% in the presence of 5 wt.% of water in the conventional mechanical stirring method. Furthermore, the excess amounts of methanol and acetone in the BDF layer after phase separation were simultaneously recovered by distilling the BDF layer at 60 °C under reduced pressure of 0.1 atm, and they were used for the next experiment. Maeda *et al.* also elucidated that the retardation of FAME formation after the glycerol formation could be explained due to the elimination of reactant methanol, which is easily dissolved into glycerol, but not due to the back reaction of the products. The co-solvent method could be recognized as a new green technology for the production of renewable biomass energy because BDF can be produced with minimum energy consumption and minimum waste emission. The optimal results from this work were applied to produce good quality of BDF from catfish oil on a pilot plant scale with a capacity of 300 L per batch. The time consumption for production of 300 L BDF from catfish oil at this pilot plant was 3 h which is shorter than the conventional method (12–20 h) [6].

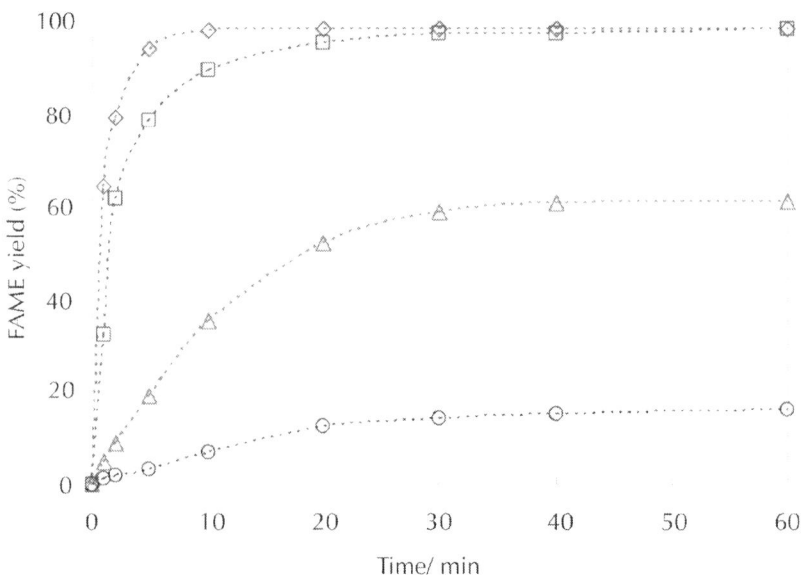

**Figure 4**: The effect of water on the formation rate of FAME (Notes: ( ◊ ) 2 wt.% water, co-solvent; ( □ ) 5 wt.% water, co-solvent; ( △ ) 2 wt.% water, mechani-

cal stirring; ( ) 5 wt.% water, mechanical stirring. The conditions: molar ratio of methanol to waste cooking oil, 4.5:1; solvent acetone to oil, 25 wt.%; KOH to oil, 0.5 wt.%; temperature, 20 °C). Reprinted with permission from [6]. Copyright 2010 Royal Society of Chemistry.

# Continuous Method Using a Gas-Liquid Reactor

A novel continuous reactor process has been developed for the production of biodiesel from fats and oils. This process was performed by atomizing the heated oil/fat and then introducing it into a reaction chamber filled with methanol and alkaline catalyst vapor in a counter current flow arrangement. The atomization process increased the oil/methanol contact area by producing micro sized droplets of 100–200 µm, and therefore increased the heat and mass transfer that is vital for a rapid reaction. In addition, the process allows the use of a very high excess of methanol since unlike the batch process methanol vapor can be recycled back to the reactor without requiring an expensive separation process and intensive energy. The transesterification of soybean oil with methanol was carried out in the continuous gas-liquid reactor with optimal conditions of NaOH 5–7 g $L^{-1}$ of methanol; methanol flow rate of 17.2 L $h^{-1}$; oil flow rate of 10 L $h^{-1}$; and temperatures 100–120 °C. Under these conditions, the conversion of triglyceride can be achieved of 94–96% [90].

## *Manganese (II) Oxide (MnO) and Titanium (II) Oxide (TiO) Catalysts*

Recently, Gombotz *et al.* have used Manganese (II) oxide (MnO) and titanium (II) oxide (TiO) as solid catalysts for both the transesterification of triglycerides and the esterification of FFA into FAME. These catalysts can be applied for low quality feedstocks containing high water content without the pretreatment steps as for the traditional process. In this study, a continuous reactor of a stainless steel tube with an inside diameter of 0.85 cm and a length of 23 cm packed with either MnO (28.1 g) or TiO (36.9 g) was used. The oil and methanol were introduced into the reactor by a HPLC pump with flow rates adjusted to provide a methanol to oil molar ratio of 6:1–30:1. A backpressure gauge was utilized on the outlet side of the column to apply a backpressure of 8.3–9.0 MPa. They produced high quality BDF (meeting ASTM specifications) from yellow grease with 15% FFA at the optimal reaction conditions: 29:1 methanol to oil mol ratio in stage 1, 15:1 methanol to oil molar ratio in stage 2, and reaction temperature 260 °C at pressure 9.0 MPa with MnO catalyst [91].

Table 5 presents comparisons of production methods and reaction conditions using various types of catalysts and oils of the yield or conversion of FAME.

# DEVELOPMENT OF NEW UTILIZATION AND REFORMING TECHNIQUES FOR GLYCERIN

When BDF is produced as an alternative to petro-based diesel fuel, a large amount of glycerol is formed as a by-product. Glycerol is currently used as an additive and a media for pharmaceuticals, cosmetics, foods, *etc.*, however, the amount of glycerol is too much to apply to such applications: the balance between the supply and demand of glycerol would break down when the industrial BDF production starts on a full scale. Therefore, it is necessary to develop new utilization and reforming techniques for glycerol. Several recent works for the development of such techniques for glycerol are described here.

**Table 5:** Summary of production methods, kind of catalysts and reaction conditions on the fatty acid methyl ester (FAME) yield

| Methods | Oils and fats | Catalysts | Reaction conditions | | | | Yield/Conversion (Y/C, %) | References |
|---|---|---|---|---|---|---|---|---|
| | | | Temperature (°C) | Molar ratio (methanol to oil) | Catalyst amount (wt.%) | Reaction time (h) | | |
| | | *Homogeneous base catalyst* | | | | | | |
| Mechanical stirring | Used frying | NaOH | 60 | 7:1 | 1.1 | 0.33 | Y = 88.8 | [10] |
| Mechanical stirring | Waste cooking | KOH | 70 | | 1 | 1 | Y = 98.2 | [92] |
| Ultrasonic irradiation | Canola, soybean | NaOH | 25 | 6:1 | 0.5 | 0.33 | Y = 98 | [76] |
| Ultrasonic irradiation | Soy bean | KOH | 40 | 6:1 | 1.5–2.2 | 0.25 | Y = 99.4 | [84] |
| Ultrasonic irradiation | Canola | KOH | 25 | 5:1 | 0.7 | 50 | Y = 99 | [9] |
| Ultrasonic irradiation | Waste cooking | KOH (two-step reaction) | 27–32 | 4:1 | 1 | 0.016 | Y= 99 | [11] |
| | | *Homogeneous acid catalyst* | | | | | | |

| Mechanical stirring | Waste cooking | H$_2$SO$_4$ | 95 | 20:1 | 4 | 20 | C > 90 | [93] |
|---|---|---|---|---|---|---|---|---|
| Mechanical stirring | Sun flower | H$_2$SO$_4$ | 65 | 30:1 | 1 | 69 | C = 90 | [94] |
| | | *Two-step: acid catalyst follow by base catalyst* | | | | | | |
| Mechanical stirring | Karanja | First-step H$_2$SO$_4$ | 60 | 6:1 | 2.2 | 1 | FFA, C = 90.6 | [95] |
| Mechanical stirring | Karanja | Second-step KOH | 60 | 8:1 | 1 | 1 | Y = 96–100 | [95] |
| Mechanical stirring | Waste cooking | First-step Fe$_3$(SO$_4$)$_3$ | 60 | 7:1 | 0.4 | 3 | Y = 81.3 | [96] |
| Mechanical stirring | Waste cooking | Second-step CaO | 60 | 7:1 | Not specified | 3 | | |
| | | *Heterogeneous base catalyst* | | | | | | |
| Mechanical stirring | Palm kernel | CaO | 60 | 30:1 | 6 | 3 | Y = 98 | [53] |
| Supercritical methanol | Sun-flower | CaO | 252 | 41:1 | 3 | 0.1 | completed | [43] |
| Mechanical stirring | Soy bean | MgO | 130 | 55:1 | 5 | 7 | Y = 60 | [50] |

| | | | | | | | | |
|---|---|---|---|---|---|---|---|---|
| Mechanical stirring | Waste cooking | $K_3PO_4$ | 60 | 6:1 | 4 | 2 | Y = 97.3 | [97] |
| | | *Heterogeneous acid catalyst* | | | | | | |
| Microwave | Yellow horn | $Cs_{2.5}H0_{0.5}PW_{12}O_{40}$ | 60 | 12:1 | 1 | 0.16 | Y = 96.22 | [98] |
| Mechanical stirring | Waste cooking | $SO_4^{2-}/ZrO_2$ | 120 | 9:1 | 3 | 4 | Y = 93.6 | [99] |
| Mechanical stirring | Soybean | $Sr(NO_3)_2/ZnO$ | 65 | 12:1 | 5 | 4 | Y = 94.7 | [100] |
| | | *Enzymatic catalyst* | | | | | | |
| Mechanical stirring | Waste edible | Novozym 435 | 30 | 3:1 | 4 | 50 | C = 90.9 | [101] |
| Mechanical stirring | Waste cooking | Rhizopus oryzae | 40 | 4:1 | 30 | 30 | Y = 88–90 | [102] |
| Mechanical stirring | grease | Pseudomonas cepacia (PS30) | 38.4 | Ethanol (6.6:1) | 13.7 | 2.47 | Y = 96 | [103] |

# Reforming of Glycerol to Produce Biofuels and Valuable Chemicals by Bioprocessing

Bioprocessing of glycerol to produce biofuels and alternative chemicals has been investigated actively [104,105,106,107,108,109,110]. Figure 5 shows examples of products synthesized from glycerol fermentation [105]. It can be seen that the formation of 1,3-propanediol, succinic acid, butanol, ethanol, formic acid, propionic acid, $H_2$ and $CO_2$ occurs during anaerobic fermentation of glycerol. Yazdani and Gonzalez reported that the maximum theoretical yield in each case from glycerol is higher than that obtained from the use of common sugars such as glucose and xylose [105].

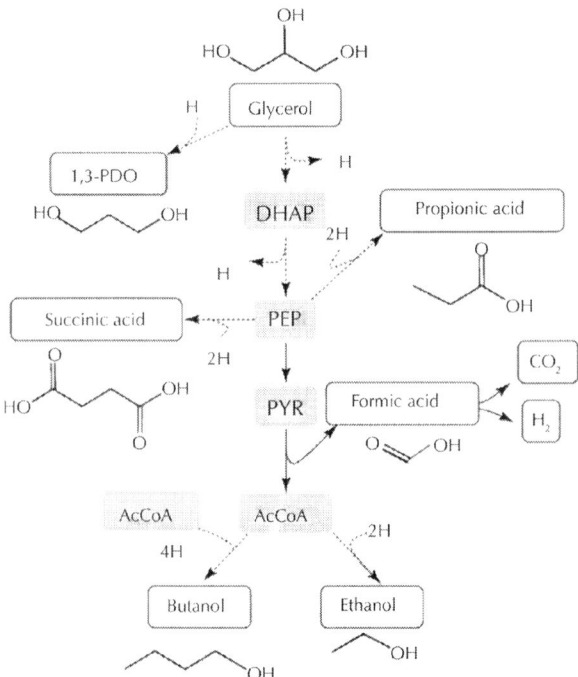

**Figure 5**: Examples of products synthesized from glycerol fermentation. Broken lines represent pathways composed of several reactions. (Abbreviations: AcCoA, acetyl-coenzyme A; DHAP, dihydroxyacetone phosphate; PEP, phosphoenolpyruvate; PYR, pyruvate; 1,3-PDO, 1,3-propanediol). Reprinted with permission from [105]. Copyright 2007 Elsevier.

1,2-propanediol can also be produced from glycerol by using metabolic engineering *Escherichia coli* [106]. Diols such as 1,3-propanediol, 1,2-propanediol, *etc.*, are useful chemicals as platform chemicals. For example, 1,3-propanediol has been used as a monomer for the synthesis of polytrimethylene terephthalate (PTT) which can be used as a fiber. It is easy to imagine the importance of PTT when we say that it is related to polyethylene terephthalate, which is well-known as PET. 1,2-propandiol can also be used in various ways as a monomer for the synthesis of polyesters and as antifreeze in breweries etc. [107]. To enhance the yield of valuable chemicals from the fermentation of glycerol, a number of attempts such as strain-based improvements and process-based improvements have been performed [105]. Trinh and Srienc investigated the conversion of glycerol to ethanol with an *Escherichia coli* strain which was designed on the basis of elementary mode analysis. They reported that the evolved strain was able to convert 40 g/L of glycerol to ethanol in 48 h with 90% of the theoretical ethanol yield [109].

# Utilization of Glycerol as a Sustainable Solvent for Green Chemistry

In chemical approaches, one of the fundamental uses of glycerol is its use as a solvent for catalysis, organic synthesis, inorganic synthesis, as well as separation and material chemistry [111,112,113]. Taking into account the properties of glycerol such as low toxicity, good biodegradability and low vapor pressure (high boiling point), glycerol has recently been shown to be an excellent sustainable solvent. For example, an advantage of the use of glycerol as a solvent is that chemical reactions can be carried out at higher temperature compared with low boiling point solvents, therefore, acceleration of the reactions or progress of different reaction pathways would be expected. As a disadvantage, the chemical reactivity of the hydroxyl groups of glycerol has to be taken into consideration. As a simple idea, glycerol can be used as a solvent instead of conventional alcohols such as methanol, ethanol, ethylene glycol, *etc.* However, we should use glycerol not only as an alternative to the conventional alcohol solvents but also as an effective solvent to enhance the rate of reactions, selectivity of reactions or yield of products. Gu *et al.* investigated an aza-Michael reaction of p-anisidine with butyl acrylate in different solvent systems under catalyst-free conditions [113]. The products were analyzed after 20 h of reaction at 100 °C. In general, aza-Michael reactions are performed in the presence of an appropriate catalyst such as Pd and Cu complexes, Lewis acids, Bronsted acids, etc., to enhance the yield of products. Gu et al. found that no reaction occurred in toluene, dimethylformamide, dimethyl sulfoxide and 1,2-dichloroethane under catalyst-free conditions, but glycerol acted as a very efficient promoting medium for this reaction (yield: about 80%). This promoting effect is due to the fact that the hydroxyl groups of glycerol are able to directly catalyze the reaction. Although water also acted as a catalyst (yield: <5%), the affinity of glycerol to p-anisidine was considered to be better than that of water to p-anisidine. In addition, it should be noted that the aza-Michael reaction proceeds effectively even in a crude glycerol solvent including about 15 wt.% of water and 5 wt.% of soap (yield: about 80%). A number of reactions which can be performed in glycerol are reviewed elsewhere [111,112].

## Utilization of Glycerol for Energy Generation

Direct alcohol fuel cells are actively being researched nowadays, because alcohols such as methanol, ethanol, ethylene glycol and glycerol have an advantage compared to hydrogen in terms of volumetric energy density. In addition, the handling of alcohols is easier for storage and transport compared to that of hydrogen.

Bianchini and Shen pointed out that unlike Pt-based electrocatalysts, Pd-based electrocatalysts would be highly active for the oxidation of a large variety of substrates in alkaline solution [114]. Since BDF is effectively synthesized in the presence of alkaline catalysts such as NaOH and KOH as seen in the previous sections, Pd-based electrocatalysts should be convenient to use without pH adjustment for crude glycerol formed from the BDF industry. Here, Pd-based electrocatalysts are briefly introduced on the basis of recent works.

Wang et al. reported the preparation of Pd/(carbonized porous anodic alumina, CPAA) electrode by the direct reduction of $PdCl_2$ with excessive $NaBH_4$ on CPAA in aqueous solution and its electrocatalytic application for alcohol oxidation [115]. Figure 6 shows the linear potential sweep curves in 1.0 M alcohol/1.0 M KOH solution at 50 mV s⁻¹. It can be seen that all alcohols can be oxidized with a Pd/CPAA electrode. They reported that the performance of Pd/CPAA for alcohol oxidation is better than that of Pd/C. Since the characteristics of Pd catalysts were not investigated by them, further examples are shown later.

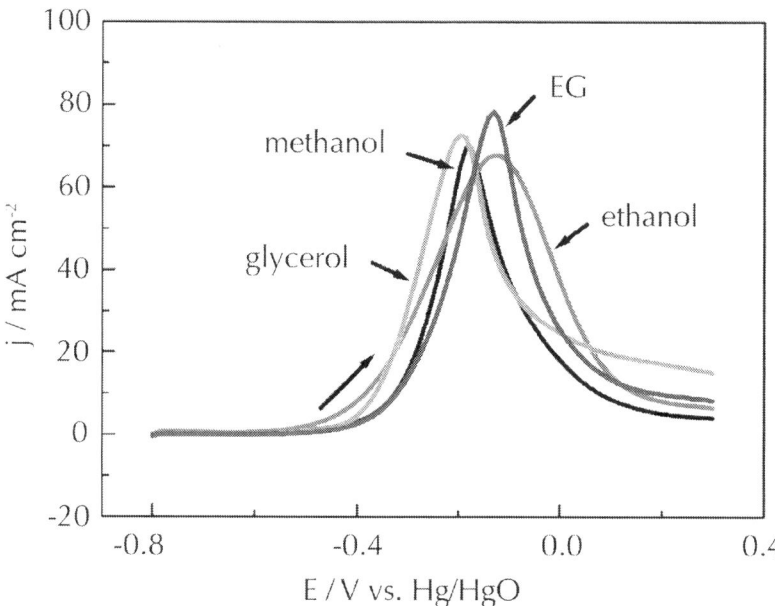

**Figure 6**: Linear potential sweep curves of the oxidation of methanol, ethanol, glycerol and ethylene glycol on the as-prepared three-dimensional Pd/ (carbonized porous anodic alumina) electrode in 1.0 M alcohol/1.0 M KOH solution, 303 K, scan rate: 50 mV s⁻¹. Reprinted with permission from [115]. Copyright 2006 Elsevier.

Figure 7 shows cyclic voltammograms of the oxidation of methanol, ethanol and glycerol on Pd nanoparticles supported on multi-walled carbon nanotubes (Pd/MWCNT) in 2 M KOH solution, where Pd/ MWCNT was synthesized by using the impregnation-reduction method. The average size of Pd nanoparticles was 4.3 nm [116]. It can be seen that Pd nanoparticles are the active catalyst for the oxidation of all alcohols investigated here. From Figure 7, the peak current density is found to be in the order of 2.8 mA/(μg-Pd) for oxidation of 5% glycerol > 2.1 mA/(μg-Pd) for oxidation of 10% ethanol > 1.1 mA/(μg-Pd) for oxidation of 10% methanol. This result shows that glycerol is the best performing fuel in spite of the lower concentration.

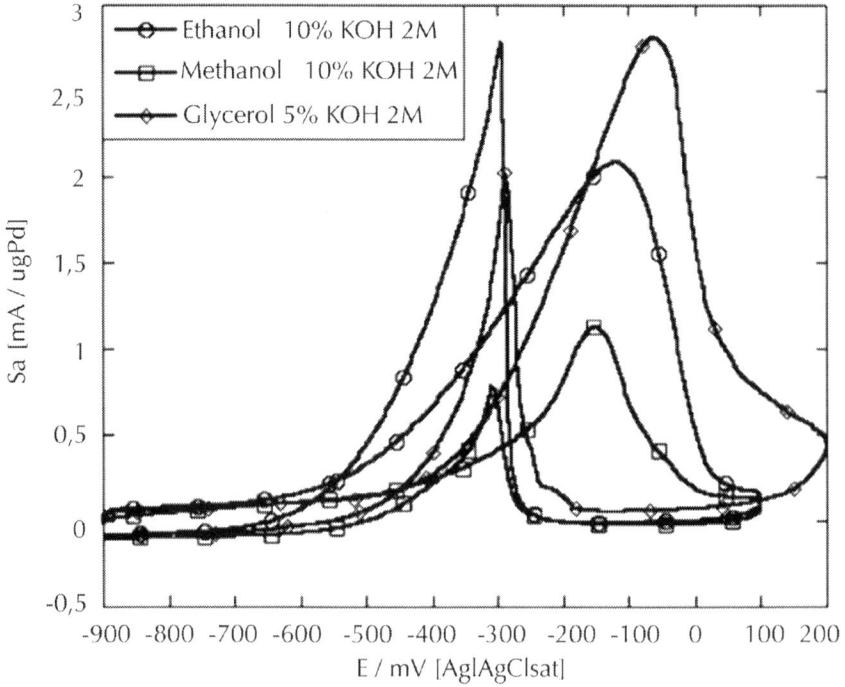

**Figure 7**: Cyclic voltammograms (at the fifth cycle) of methanol, ethanol and glycerol oxidation on a Pd/(multi-walled carbon nanotubes) electrode in 2 M KOH solution. Pd loading: 17 µg cm$^{-2}$. Scan rate: 50 mVs$^{-1}$. Average size of Pd: 4.3 nm. Reprinted with permission from [116]. Copyright 2009 Elsevier.

The surface modification of foreign atoms to Pd or Pt is suggested to enhance and improve their catalytic activity for alcohol oxidation. Simões *et al.* investigated the effects of modification of Bi to Pd or Pt on glycerol oxidation, where Pt, Pd, Pd$_{0.9}$Bi$_{0.1}$, Pt$_{0.9}$Bi$_{0.1}$ and Pd$_{0.45}$Pt$_{0.45}$Bi$_{0.1}$ nanoparticles were synthesized by the "water-in-oil" microemulsion method [117]. The average size of the particles prepared was 4.0 nm for Pd, 5.3 nm for Pt, 5.2 nm for Pd$_{0.9}$Bi$_{0.1}$, 4.7 nm for Pt$_{0.9}$Bi$_{0.1}$, and 4.5 nm for Pd$_{0.45}$Pt$_{0.45}$Bi$_{0.1}$, respectively. Based on analyzing the onset potential of the oxidation wave, it was found that the catalytic activity for glycerol oxidation was in the order of Pd/C < Pt/C = Pd$_{0.9}$Bi$_{0.1}$/C

$< Pt_{0.9}Bi_{0.1}/C = Pd_{0.45}Pt_{0.45}Bi_{0.1}/C$. The enhancement of the catalytic activity by adding Bi on Pd and/or Pt was suggested to be due to the changes in the electronic interactions between the reactant and the active sites of the catalyst, which are induced by the bifunctional effect and/or by the ensemble effect. The products formed during glycerol oxidation with $Pd_{0.9}Bi_{0.1}/C$, $Pt_{0.9}Bi_{0.1}/C$ and $Pd_{0.45}Pt_{0.45}Bi_{0.1}/C$ catalysts were tartronate, mesoxalate, oxalate and formate ions which were confirmed by HPLC combined with chronoamperometry experiments. This oxidation mechanism was almost the same as previous reports with other electrocatalysts [114].

The researches of direct methanol or ethanol fuel cells are advancing quickly compared with those of direct glycerol fuel cells. It is probable that similar catalysts for the oxidation of methanol and ethanol are effective for the oxidation of glycerol.

# Reforming of Glycerol to Valuable Chemicals by Catalysis

The reforming of glycerol is actively being researched by catalysis. Zhou *et al.* summarized the comprehensive review about catalytic conversion of glycerol to valuable chemicals in detail [118]. To convert glycerol into valuable chemicals, oxidation, hydrogenolysis, dehydration, pyrolysis/gasification, transesterification/esterification, etherification, oligomerization/polymerization, chlorination and carboxylation of glycerol have been investigated under various experimental conditions in the presence of catalysts. Here, several recent works are briefly introduced.

In the case of selective oxidation of glycerol, the formation of various products such as dihydroxyacetone, hydroxypyruvic acid, *etc.*, has been reported to occur. Takagaki *et al.* reported selective oxidation of glycerol to glycolic acid in water with molecular oxygen by use of hydrotalcite-supported gold nanoparticle catalysts [119]. They found that a high yield (53%) of glycolic acid was obtained at 293 K compared to 333 K. This is due to the fact that the basicity of hydrotalcite acts not only as promoter by proton abstraction of alcohol but also as *in situ* generator of hydrogen peroxide.

In hydrogenolysis of glycerol, 1,2-propanediol, 1,3-propanediol and ethylene glycol can be synthesized selectively. Wu *et al.* reported the synthesis of 1,2-propanediol from hydrogenolysis of glycerol over a Cu-Ru/carbon nanotube catalyst [120]. The conversions of glycerol and selectivity for the formation of 1,2-propanediol were 99.8% and 86.5%, respectively. Shimao et al. reported the promoting effect of Re addition to Rh/SiO$_2$ on glycerol hydrogenolysis [121]. They found that the modification of ReO$_x$ to Rh enhanced the activity of glycerol hydrogenolysis and the formation of 1,3-propanediol became more favorable on the Rh-ReO$_x$/SiO$_2$. Ueda et al. reported that the formation of ethylene glycol in glycerol hydrogenolysis was enhanced over Pt-modified Ni catalyst, where the conversion of glycerol to ethylene glycol was suggested to occur via retro-aldol reaction of glyceraldehyde [122].

The chlorination of glycerol has been investigated to produce dichloropropanol [123,124,125,126] which can be used as an intermediate for epichlorohydrin. In addition, the etherification of glycerol with isobutylene has been investigated to produce an oxygenate additive which can be used as an ignition accelerator and octane booster [127].

A number of papers have reported the formation of gaseous products form glycerol reforming. Vaidya and Rodrigues reviewed H$_2$ production from glycerol reforming over Ni, Pt and Ru catalysts [128]. The synthesis of H$_2$ and CO from glycerol has also been investigated over Pt-based catalysts [129]. It is important to develop an effective catalytic process to transform glycerol to various useful chemicals in the future.

# CONCLUSIONS

Biodiesel is a renewable and alternative fuel to petro diesel fuel. In addition, biodiesel is environmental friendly due to its easy biodegradability, non-toxicity, being primarily free of sulfur and aromatics and containing oxygen in its structure resulting in production of more tolerable exhaust gas emissions than conventional fossil diesel, despite providing similar levels of fuel efficiency. Currently, biodiesel is produced thank to esterification and transesterification reactions from edible and non-edible vegetable oils or animal fats with primary alcohols in the presence of an acid- or base-catalyst. Several catalysts such as homogeneous acid/base, heterogeneous acid/

base, enzymes, *etc.* have been studied and applied to the synthesis of biodiesel. However, in commercial production, a homogeneous alkaline catalyst transesterification is predominately used for good quality oils containing a low content of FFA because the base alkaline catalyst gives a high FAME yield in a short reaction time and the reaction can be carried out in simple equipment. In contrast, with poor quality raw oils containing a high amount of FFA, a strong sulfuric acid catalyst esterification used as a pre-treatment step followed by an alkaline catalyst transesterification is the most popular way to produce biodiesel.

Currently, the mechanical stirring method with a batch reactor is the conventional method for biodiesel production on the industrial scale, because this method is simple and cheap. However, the production process has long reaction times and separation of crude BDF from the reaction mixture, and the reaction is performed at relatively high temperature with a base-catalyst resulting in soap formation. To solve these disadvantages, the ultrasonic irradiation and co-solvent methods have been developed and applied for the production of biodiesel on the industrial scale. With these innovative methods, the reaction can be conducted at ambient temperature with shorter reaction times and reduced raw material consumption. Combination of these new methods with solid catalysts will give green technologies for production of biodiesel in the near future.

In addition, new utilization technologies for glycerol must be developed to reduce the amount of glycerol waste. While various technologies such as "reforming of glycerol to produce biofuels and valuable chemicals by bioprocessing or catalysis", "utilization of glycerol as a sustainable solvent for green chemistry" and "utilization of glycerol for energy generation", are being actively studied by a number of researchers, the catalysis process could become one of the most important processes to reform glycerol to useful chemicals in the future.

# ACKNOWLEDGMENTS

We acknowledge the support from Science and Technology Research Partnership for Sustainable Development (SATREPS, Project: Multi-beneficial Measure for the Mitigation of Climate Change by the

Integrated Utilization of Biomass Energy in Vietnam and Indochina countries), JST-JICA, Japan.

# REFERENCES

1.  Demirbas, A. Political, economic and environmental impacts of biofuels: A review. *Appl. Energy* 2009, *86*, S108–S117.
2.  Georgogianni, K.G.; Kontominas, M.G.; Tegou, E.; Avlonitis, D.; Gergis, V. Biodiesel production: Reaction and process parameters of alkali-catalyzed transesterification of waste frying oils. *Energy Fuels* 2007, *21*, 3023–3027.
3.  Zah, R.; Ruddy, T.F. International trade in biofuels: An introduction to the special issue. *J. Clean. Prod.* 2009, *17*, S1–S3.
4.  Eisentraut, A. *Technology Roadmap Biofuels for Transport*; International Energy Agency: Paris, France, 2011; p. 12.
5.  Balat, M.; Balat, H. A critical review of bio-diesel as a vehicular fuel. *Energy Convers. Manag.* 2008, *49*, 2727–2741.
6.  Maeda, Y.; Thanh, L.T.; Imamura, K.; Izutani, K.; Okitsu, K.; Boi, L.V.; Lan, P.N.; Tuan, N.C.; Yoo, Y.E.; Takenaka, N. New technology for the production of biodiesel fuel. *Green Chem.* 2010, *13*, 1124–1128.
7.  Knothe, G.; Gerpen, J.V.; Krahl, J. *The Biodiesel Handbook*; AOCS Press: Champaign, IL, USA, 2005; pp. 34, 35, 164, 269, 270-274.
8.  Jain, S.; Sharma, M.P. Biodiesel production from *Jatropha curcas* oil. *Renew. Sustain. Energy Rev.* 2010, *14*, 3140–3147.
9.  Thanh, L.T.; Okitsu, K.; Sadanaga, Y.; Takenaka, N.; Yasuaki Maeda, Y.; Bandow, H. Ultrasound-assisted production of biodiesel fuel from vegetable oils in a small scale circulation process. *Bioresour. Technol.* 2010, *101*, 639–645.
10. Leung, D.Y.C.; Guo, Y. Transesterification of neat and used frying oil: Optimization for biodiesel production. *Fuel Process. Technol.* 2006, *87*, 883–890.
11. Thanh, L.T.; Okitsu, K.; Sadanaga, Y.; Takenaka, N.; Maeda, Y.; Bandow, H. A two-step continuous ultrasound assisted production of biodiesel fuel from waste cooking oils: A practical

and economical approach to produce high quality biodiesel fuel. *Bioresour. Technol.* 2010, *101*, 5394–5401.

12. Ghanei, R.; Moradi, G.R.; TaherpourKalantari, R.; Arjmandzadeh, E. Variation of physical properties during transesterification of sunflower oil to biodiesel as an approach to predict reaction progress. *Fuel Process. Technol.* 2011, *92*, 1593–1598.

13. Vujicic, D.; Comic, D.; Zarubica, A.; Micic, R.; Boskovic, G. Kinetics of biodiesel synthesis from sunflower oil over CaO heterogeneous catalyst. *Fuel* 2010, *89*, 2054–2061.

14. Corro, G.; Tellez, N.; Ayala, E.; Marinez-Ayala, A. Two-step biodiesel production from *Jatropha curcas* crude oil using SiO2·HF solid catalyst for FFA esterification step. *Fuel* 2010, *89*, 2815–2821.

15. Rashid, U.; Anwar, F. Production of biodiesel through base-catalyzed transesterification of safflower oil using an optimized protocol. *Energy Fuels* 2008, *22*, 1306–1312.

16. Hayyan, A.; Alam, M.Z.; Mirghani, M.E.S.; Kabbashi, N.A.; Hakimi, N.I.N.M.; Siran, Y.M.; Tahiruddin, S. Reduction of high content of free fatty acid in sludge palm oil via acid catalyst for biodiesel production. *Fuel Process. Technol.* 2011, *92*, 920–924.

17. Crabbe, E.; Nolasco-Hipolito, C.; Kobayashi, G.; Sonomoto, K.; Ishizaki, A. Biodiesel production from crude palm oil and evaluation of butanol extraction and fuel properties. *Process Biochem.* 2001, *37*, 65–71.

18. Shu, Q.; Zhang, Q.; Xu, G.; Nawaz, Z.; Wang, D.; Wang, J. Synthesis of biodiesel from cottonseed oil and methanol using a carbon-based solid acid catalyst. *Fuel Process. Technol.* 2009, *90*, 1002–1008.

19. Qian, J.; Yun, Z.; Shi, H. Cogeneration of biodiesel and nontoxic cottonseed meal from cottonseed processed by two-phase solvent extraction. *Energy Convers. Manag.* 2010, *51*, 2750–2756.

20. Bi, Y.; Ding, D.; Wang, D. Low-melting-point biodiesel derived from corn oil via urea complexation. *Bioresour. Technol.* 2010, *101*, 1220–1226.

21. Moreau, R.A.; Powell, M.J.; Hicks, K.B. Extraction and quantitative analysis of oil from commercial corn fiber. *J. Agric. Food Chem.* 1996, *44*, 2149–2154.

22.  Kumar, D.; Kumar, G.; Poonam; Singh, C.P. Fast, easy ethanolysis of coconut oil for biodiesel production assisted by ultrasonication. *Ultrason. Sonochem.* 2010, *17*, 555–559.

23.  Nakpong, P.; Wootthikanokkhan, S. High free fatty acid coconut oil as a potential feedstock for biodiesel production in Thailand. *Renew. Energy* 2010, *35*, 1682–1687.

24.  Kouzu, M.; Kasuno, T.; Tajika, M.; Sugimoto, Y.; Yamanaka, S.; Hidaka, J. Calcium oxide as a solid base catalyst for transesterification of soybean oil and its application to biodiesel production. *Fuel* 2008, *87*, 2798–2806.

25.  Trentin, C.M.; Lima, A.P.; Alkimim, I.P.; da Silva, C.; de Castilhos, F.; Mazutti, M.A.; Oliveira, J.V. Continuous production of soybean biodiesel with compressed ethanol in a microtube reactor using carbon dioxide as co-solvent. *Fuel Process. Technol.* 2011, *92*, 952–958.

26.  Encinar, J.M.; Sanchez, N.; Martinez, G.; Garcia, L. Study of biodiesel production from animal fats with high free fatty acid content. *Bioresour. Technol.* 2011, *102*, 10907–10914.

27.  Cheng, L.H.; Yen, S.Y.; Su, L.S.; Chen, J. Study on membrane reactors for biodiesel production by phase behaviors of canola oil methanolysis in batch reactors. *Bioresour. Technol.* 2010, *101*, 6663–6668.

28.  Dizge, N.; Keskinler, B. Enzymatic production of biodiesel from canola oil using immobilized lipase. *Biomass Bioenergy* 2008, *32*, 1274–1278.

29.  Phan, A.N.; Phan, T.M. Biodiesel production from waste cooking oils. *Fuel* 2008, *87*, 3490–3496. [Google Scholar] [CrossRef]

30.  Freedman, B.; Butterfield, R.O.; Pryde, E.H. Transesterification kinetics of soybean oil. *J. Am. Oil Chem. Soc.* 1986, *63*, 1375–1380. [Google Scholar] [CrossRef]

31.  Darnoko, D.; Cheryan, M. Kinetics of palm oil transesterification in a batch reactor. *J. Am. Oil Chem. Soc.* 2000, *77*, 1263–1267.

32.  Lee, D.W.; Park, Y.M.; Lee, K.Y. Heterogeneous base catalysts for transesterification in biodiesel synthesis. *Catal. Surv. Asia* 2009, *13*, 63–67.

33. Meher, L.C.; Sagar, D.V.; Naik, S.N. Technical aspects of biodiesel production by tranesterification—a review. *Renew. Sustain. Energy Rev.* 2006, *10*, 248–268.

34. Georgogianni, K.G.; Katsoulidis, A.P.; Pomonis, P.J.; Kontominas, M.G. Transesterification of soybean frying oil to biodiesel using heterogeneous catalysts. *Fuel Process. Technol.* 2009, *90*, 671–676.

35. Ilham, Z.; Saka, S. Dimethyl carbonate as potential reactant in non-catalytic biodiesel production by supercritical method. *Bioresour. Technol.* 2009, *100*, 1793–1796.

36. Ji, J.; Wang, J.; Li, Y.; Yu, Y.; Xu, Z. Preparation of biodiesel with the help of ultrasonic and hydrodynamic cavitation. *Ultrasonics* 2006, *44*, 411–414.

37. Meher, L.C.; Dharmagadda, V.S.S.; Naik, S.N. Optimization of alkaline-catalyzed transesterification of *Pongamia pinnata* oil for production of biodiesel. *Bioresour. Technol.* 2006, *97*, 1392–1397.

38. Noureddini, H.; Harkey, D.; Medikonduru, V. A continuous process for the conversion of vegetable oils into methyl esters of fatty acids. *J. Am. Oil Chem. Soc.* 1998, *75*, 1775–1783.

39. Ramachandran, K.B.; Al-Zuhair, S.; Fong, C.S.; Gak, C.W. Kinetic study on hydrolysis of oils by lipase with ultrasonic emulsification. *Biochem.Eng. J.* 2006, *32*, 19–24.

40. Helwani, Z.; Othman, M.R.; Aziz, N.; Fernando, W.J.N.; Kim, J. Technologies for production of biodiesel focusing on green catalytic techniques: A review. *Fuel Process. Technol.* 2009, *90*, 1502–1514.

41. Schuchardta, R.; Serchelia, R.; Vargas, R.M. Transesterification of vegetable oils: A review. *J. Braz. Chem. Soc.* 1998, *9*, 199–210.

42. Sivakumar, P.; Anbarasu, K.; Renganathan, S. Bio-diesel production by alkaline catalyzed transesterification of dairy waste scum. *Fuel* 2011, *90*, 147–151.

43. Canakci, M.; Gerpen, J.V. A pilot plant to produce biodiesel from high free fatty acid feedstocks. *Trans. Autom. Sci. Eng.* 2003, *46*, 945–955.

44.  Ma, F.; Clements, L.D.; Hanna, M.A. Biodiesel from animal fat. Ancillary studies on transesterification of beef tallow. *Ind. Eng. Chem. Res.* 1998, *37*, 3768–3771.

45.  Canakci, M.; Gerpen, J.V. Biodiesel production via acid-catalyst. *Trans. Autom. Sci. Eng.* 1999, *42*, 1203–1210. [Google Scholar]

46.  Patil, P.D.; Gude, V.G.; Deng, S. Biodiesel production from *Jatropha curcas*, waste cooking, and camelina Sativa. *Ind. Eng. Chem. Res.* 2009, *48*, 10850–10856.

47.  Arzamendi, G.; Campoa, I.; Arguinarena, E.; Sanchez, M.; Montes, M.; Gandia, L.M. Synthesis of biodiesel with heterogeneous NaOH/alumina catalysts: comparison with homogeneous NaOH. *Chem. Eng. J.* 2007, *134*, 123–130.

48.  Perego, C.; Bosetti, A. Biomass to fuels: The role of zeolite and mesoporous materials. *Microporous Mesoporous Mater.* 2011, *144*, 28–39.

49.  Demirbas, A. Biodiesel from sunflower oil in supercritical methanol with calcium oxide. *Energy Convers. Manag.* 2007, *48*, 937–941.

50.  Chouhan, P.S.; Sarma, A.K. Modern heterogeneous catalysts for biodiesel production: A comprehensive review. *Renew. Sustain. Energy Rev.* 2011, *15*, 4378–4399.

51.  Antunes, W.M.; Veloso, C.O.; Henriques, C.A. Transesterification of soybean oil with methanol catalyzed by basic solids. *Catal. Today* 2008, *133–135*, 548–554.

52.  Verziu, M.; Cojocaru, B.; Hu, J.; Richards, R.; Ciuculescu, C.; Filip, P.; Parvulescu, V.I. Sunflower and rapeseed oil transesterification to biodiesel over different nanocrytalline MgO catalysts. *Green Chem.* 2008, *10*, 373–381.

53.  Sharma, Y.C.; Signh, B.; Korstad, J. Latest developments on application of heterogeneous basic catalysts for an efficient and eco friendly synthesis of biodiesel: A review. *Fuel* 2011, *90*, 1309–1324.

54.  Ngamcharussrivichai, C.; Nunthasanti, P.; Tanachai, S.; Bunyakiat, K. Biodiesel production through transesterification over natural calciums. *Fuel Process. Technol.* 2010, *91*, 1409–1415.

55.  Huaping, Z.; Zongbin, W.; Yuanxiao, C.; Ping, Z.; Shije, D.; Xiaohua, L.; Zongqian, M. Preparation of biodiesel catalyzed by

solid super base of calcium hydroxide and its refining process. *Chin. J. Catal.* 2006, *27*, 391–396.

56. Grandos, M.L.; Poves, M.D.; Alonso, D.; Miriscal, R.; Galisteo, F.C. Biodiesel from sunflower oil by using activated calcium oxide. *Appl. Catal. B* 2007, *73*, 317–326.

57. Omar, W.N.N.W.; Amin, N.A.S. Biodiesel production from waste cooking oil over alkaline modified zirconia catalyst. *Fuel Process. Technol.* 2011, *92*, 397–2405.

58. Guan, G.; Kusakabe, K.; Yamasaki, S. Tri-potassium phosphate as a solid catalyst for biodiesel production from waste cooking oil. *Fuel Process. Technol.* 2009, *90*, 520–524.

59. Jacobson, K.; Gopinath, R.; Meher, L.C.; Dalai, A.K. Solid acid catalyzed biodiesel production from waste cooking oil. *Appl. Catal.B* 2008, *85*, 86–91.

60. Wen, Z.; Yu, X.; Tu, S.T.; Yan, J.; Dahlquist, E. Biodiesel production from waste cooking oil catalyzed by $TiO_2$–MgO mixed oxides. *Bioresour.Technol.* 2010, *101*, 9570–9576.

61. Lopez, D.E.; Goodwin, J.G.; Bruce, J.D.A. Transesterification of triacetin with methanol on Nafion-acid resins. *J. Catal.* 2007, *245*, 381–391.

62. Merelo, J.A.; Iglesias, J.; Morales, G. Heterogeneous acid catalysts for biodiesel production: current status and future challenges. *Green Chem.* 2009, *11*, 1285–1308.

63. Russbueldt, B.M.E.; Hoelderich, W.F. New sulfonic ion exchange resins for preesterification of different oils and fats with high content of free fatty acid. *Appl. Catal. A* 2009, *362*, 47–57.

64. Tesser, R.; Serio, M.D.; Guida, M.; Nastasi, M.; Santacesaria, E. Kinetics of oleic acid esterification with methanol in the presence of triglycerides. *Ind. Eng. Chem. Res.* 2005, *44*, 7978–7982.

65. Melero, J.A.; Bautista, L.F.; Morales, G.; Iglesias, J.; Briones, D. Biodiesel production with heterogeneous sulfonic acid-functionalized mesostructured catalysts. *Energy Fuels* 2009, *23*, 539–547.

66. Toda, M.; Takagaki, A.; Okamura, M.; Kondo, J.N.; Hayashi, S.; Domen, K.; Hara, M. Biodiesel made with sugar catalyst. *Nature* 2005, *438*, 178.

67. Hara, M. Biomass conversion by a solid acid catalyst. *Energy Environ. Sci.* 2010, *3*, 601–607.

68. Dehkhoha, A.M.; West, A.H.; Ellis, N. Biochar based solid acid catalyst for biobiodiesel production. *Appl. Catal. A* 2010, *382*, 197–204.

69. Zong, M.H.; Duan, Z.Q.; Lou, W.Y.; Smith, T.J.; Wu, H. Preparation of a sugar catalyst and its use for highly efficient production of biodiesel. *Green Chem.* 2007, *9*, 434–437.

70. Kotwal, M.S.; Niphadkar, P.S.; Deshpadkar, S.S.; Bokade, V.V.; Joshi, P.N. Transesterification of sunflower oil catalyzed by flyash-based solid catalysts. *Fuel* 2009, *88*, 1773–1778.

71. Moreira, A.B.R.; Perez, V.H.; Zanin, G.M.; Castro, H.F. Biodiesel Synthesis by Enzymatic Transesterification of Palm Oil with Ethanol Using Lipases from Several Sources Immobilized on Silica–PVA Composite. *Energy Fuels* 2007, *21*, 3689–3694.

72. Ha, S.H.; Lan, M.N.; Lee, S.H.; Hwang, S.M.; Koo, Y.M. Lipase-catalyzed biodiesel production from soybean oil in ionic liquids. *Enzym. Microb. Technol.* 2007, *41*, 480–483.

73. Modi, M.K.; Reddy, J.R.C.; Roa, B.V.S.K.; Prasad, R.B.N. Lipase-mediated conversion of vegetable oils into biodiesel using ethyl acetate as acyl acceptor. *Bioresour. Technol.* 2007, *98*, 1260–1264.

74. Shimada, Y.; Watanabe, Y.; Sugihara, A.; Tominaga, Y. Enzymatic alcoholysis for biodiesel fuel production and application of the reaction to oil processing. *J. Mol. Catal. B Enzym.* 2002, *17*, 133–142.

75. Nelson, L.A.; Foglia, T.A.; Marmer, W.N. Lipase-catalyzed production of biodiesel. *J. Am. Oil Chem. Soc.* 1996, *73*, 1191–1194.

76. Stavarache, C.; Vinatoru, M.; Nishimura, R.; Maeda, Y. Conversion of vegetable oil to biodiesel using ultrasonic irradiation. *Chem. Lett.* 2003, *32*, 716–717.

77. Stavarache, C.; Vinatoru, M.; Maeda, Y. Ultrasonic *versus* silent methylation of vegetable oils. *Ultrason. Sonochem.* 2006, *13*, 401–407.

78. Thanh, L.T.; Okitsu, K.; Sadanaga, Y.; Takenaka, N.; Bandow, H. Biodiesel production from virgin and waste oils using ultrasonic

reator in pilot scale. *Proc. Symp. Ultrason. Electron.* 2008, *29*, 395–396.

79.   Hanh, H.D.; Dong, N.T.; Stavarache, C.; Okitsu, K.; Maeda, Y.; Nishimura, R. Methanolysis of triolein by low frequency ultrasonic irradiation. *Energy Convers. Manag.* 2008, *49*, 276–280.

80.   Georgogianni, K.G.; Kontominas, M.G.; Pomonis, P.J.; Avlonitis, D.; Gergis, V. Conventional and *in situ* transesterification of sunflower seed oil for the production of biodiesel. *Fuel Process. Technol.* 2008, *89*, 503–509.

81.   Gogate, P.R.; Kabadi, A.M. A review of applications of cavitation in biochemical engineering/biotechnology. *Biochem. Eng. J.* 2009, *44*, 60–72.

82.   Stavarache, C.; Vinatoru, M.; Maeda, Y.; Bandow, H. Ultrasonically driven continuous process for vegetable oil transesterification. *Ultrason. Sonochem.* 2007, *14*, 413–417.

83.   Hanh, H.D.; Dong, N.T.; Okitsu, K.; Nishimura, R.; Maeda, Y. Biodiesel production by esterification of oleic acid with short-chain alcohols under ultrasonic irradiation condition. *Renew. Energy* 2009, *34*, 780–783.

84.   Clucci, J.A.; Borrero, E.E.; Alape, F. Biodiesel from an alkaline transesterification reaction of soybean oil using ultrasonic mixing. *J. Am. Oil Chem. Soc.* 2005, *82*, 525–530.

85.   Mootabadi, H.; Salamatinia, B.; Bhatia, S.; Abdullah, A.Z. Ultrasonic-assisted biodiesel production process from palm oil using alkaline earth metal oxides as the heterogeneous catalysts. *Fuel* 2010, *89*, 1818–1825.

86.   Stavarache, C.; Vinatoru, M.; Nishimura, R.; Maeda, Y. Fatty acids methyl esters from vegetable oil by means of ultrasonic energy. *Ultrason. Sonochem.* 2005, *12*, 367–372.

87.   Bunkyakiat, K.; Makmee, S.; Sawangkeaw, R.; Ngamprasertsith, S. Continuous production of biodiesel via transesterification from vegetable oils in supercritical methanol. *Energy Fuels* 2006, *20*, 812–817.

88.   Saka, S.; Kusdiana, D. Biodiesel fuel from rapeseed oil as prepared in supercritical methanol. *Fuel* 2001, *80*, 225–231.

89.   Demirbas, A. Biodiesel production from vegetable oils by supercritical methanol. *J. Sci. Ind. Res.* 2005, *64*, 858–865.

90. Behzadi, S.; Farid, M.M. Production of biodiesel using a continuous gas-liquid reactor. *Bioresour. Technol.* 2009, *100*, 683–689.

91. Gombotz, K.; Parette, R.; Austic, G.; Kannan, D.; Matson, J.V. MnO and TiO solid catalysts with low-grade feedstocks for biodiesel production. *Fuel* 2012, *92*, 9–15.

92. Agarwal, M.; Chauhan, G.; Chaurasia, S.P.; Singh, K. Study of catalytic behavior of KOH as homogeneous and heterogeneous catalyst for biodiesel production. *J. Taiwan Inst. Chem. Eng.* 2012, *43*, 89–94.

93. Wang, Y.; Ou, S.; Liu, P.; Xue, F.; Tang, S. Comparison of two different processes to synthesize biodiesel by waste cooking oil. *J. Mol. Catal. A Chem.* 2006, *252*, 107–112.

94. Freedman, B.; Pryde, E.H.; Mounts, T.L. Variables affecting the yields of fatty esters from transesterified vegetable oils. *J. Am. Oil Chem. Soc.* 1984, *61*, 1638–1643.

95. Sharma, Y.C.; Singh, B.; Kortad, J. High yield and conversion of biodiesel from a Nonedible feekstock (*Pongamia pinnata*). *J. Agric. Food. Chem.* 2010, *58*, 242–247.

96. Omar, W.; Nordin, N.; Mohamed, M.; Amin, N.A.S. A two-step biodiesel production from waste cooking oil: optimization of pre-treatment step. *J. Appl. Sci.* 2009, *9*, 3098–4103.

97. Guan, G.; Kusakabe, K.; Yamasaki, S. Tri-potassium phosphate as a solid catalyst for biodiesel production from waste cooking oil. *Fuel Process. Technol.* 2009, *90*, 520–524.

98. Zhang, S.; Zu, Y.-G.; Fu, Y.-J.; Luo, M.; Zhang, D.-Y.; Efferth, T. Rapid microwave-assited transeterification of yellow horn oil to biodiesel using a heterogeneuous solid catalyst. *Bioresour. Technol.* 2010, *101*, 931–936.

99. Fu, B.; Gao, L.; Niu, L.; Wei, R.; Xiao, G. Biodiesel from waste cooking oil via heterogeneous superacid catalyst $SO_4^{2-}/ZrO_2$. *Energy Fuels* 2009, *23*, 569–572.

100. Yang, Z.; Xie, W. Soybean oil transesterification over zinc oxide modified with alkaline earth metals. *Fuel process. Technol.* 2007, *88*, 631–638.

101. Watanabe, Y.; Shimada, Y.; Sugihar, A.; Tominaga, Y. Enzymatic conversion of waste edible oil to biodiesel fuel in a fixed-bed bioreactor. *J. Am. Chem. Soc.* 2001, *78*, 703–707.

102. Chen, G.; Ying, M.; Li, W. Enzymatic conversion of waste cooking oils into alternative fuel-biodiesel. *Appl. Biochem. Biotechnol.* 2006, *132*, 911–921.

103. Wu, W.H.; Foglia, T.A.; Marmer, W.N.; Phillips, J.G. Optimizing production of ethyl esters of grease using 95% ethanol by response surface methodology. *J. Am. Oil Chem. Soc.* 1999, *76*, 517–521.

104. Coombs, A. Glycerin bioprocessing goes green. *Nat. Biotechnol.* 2007, *25*, 953–954.

105. Yazdani, S.S.; Gonzalez, R. Anaerobic fermentation of glycerol: A path to economic viability for the biofuels industry. *Curr. Opin. Biotechnol.* 2007, *18*, 213–219.

106. Clomburg, J.M.; Gonzalez, R. Metabolic engineering of *Escherichia coli* for the production of 1,2-propanediol from glycerol. *Biotechnol. Bioeng.* 2011, *108*, 867–879.

107. Zeng, A.-P.; Sabra, W. Microbial production of diols as platform chemicals: recent progresses. *Curr. Opin. Biotechnol.* 2011, *22*, 749–757.

108. Ito, T.; Nakashimada, Y.; Senba, K.; Matsui, T.; Nishio, N. Hydrogen and ethanol production from glycerol-containing wastes discharged after biodiesel manufacturing process. *J. Biosci. Bioeng.* 2005, *100*, 260–265.

109. Trinh, C.T.; Srienc, F.; Choi, W.J.; Hartono, M.R.; Chan, W.H.; Yeo, S.S. Metabolic engineering of *Escherichia coli* for Efficient conversion of glycerol to ethanol. *Appl. Environ. Microbiol.* 2009, *75*, 6696–6705.

110. Choi, W.J.; Hartono, M.R.; Chan, W.H.; Yeo, S.S. Ethanol production from biodiesel-derived crude glycerol by newly isolated *Kluyvera cryocrescens*. *Appl. Microbiol. Biotechnol.* 2011, *89*, 1255–1264.

111. Gu, Y.; Jerome, F. Glycerol as a sustainable solvent for green chemistry. *Green Chem.* 2010, *12*, 1127–1138.

112. Diaz-Alvarez, A.E.; Francos, J.; Lastra-Barreira, B.; Crochet, P.; Cadierno, V. Glycerol and derived solvents: new sustainable

reaction media for organic synthesis. *Chem. Commun.* 2011, *47*, 6208–6227.

113. Gu, Y.; Barrault, J.; Jerome, F. Glycerol as an efficient promoting medium for organic reactions. *Adv. Synth. Catal.* 2008, *350*, 2007–2012.

114. Bianchini, C.; Shen, K. Palladium-based electrocatalysts for alcohol oxidation in half cells and in direct alcohol fuel cells. *Chem. Rev.* 2009, *109*, 4183–4206.

115. Wang, Z.; Hu, F.; Shen, P.K. Carbonized porous anodic alumina as electrocatalyst support for alcohol oxidation. *Electrochem. Commun.* 2006, *8*, 1764–1768.

116. Bambagioni, C.; Bianchini, A.; Marchionni, J.; Filippi, F.; Vizzaa, J.; Teddy, P.; Serp, M.; Zhiani, M. Pd and Pt–Ru anode electrocatalysts supported on multi-walled carbon nanotubes and their use in passive and active direct alcohol fuel cells with an anion-exchange membrane (alcohol = methanol, ethanol, glycerol). *J. Power Sour.* 2009, *190*, 241–251.

117. Simões, M.; Baranton, S.; Coutanceau, C. Enhancement of catalytic properties for glycerol electrooxidation on Pt and Pd nanoparticles induced by Bi surface modification. *Appl. Catal. B* 2011, *110*, 40–49.

118. Zhou, C.H.; Beltramini, J.N.; Fan, Y.X.; Lu, G.Q. Chemoselective catalytic conversion of glycerol as a biorenewable source to valuable commodity chemicals. *Chem. Soc. Rev.* 2008, *37*, 527–549.

119. Takagaki, A.; Tsuji, A.; Nishimura, S.; Ebitani, K. Genesis of catalytically active gold nanoparticles supported on hydrotalcite for base-free selective oxidation of glycerol in water with molecular oxygen. *Chem. Lett.* 2011, *40*, 150–152.

120. Wu, Z.; Mao, Y.; Wang, X.; Zhang, M. Preparation of a Cu-Ru/carbon nanotube catalyst for hydrogenolysis of glycerol to 1,2-propanediol via hydrogen spillover. *Green Chem.* 2011, *13*, 1311–1316.

121. Shimao, A.; Koso, S.; Ueda, N.; Shinmi, Y.; Furikado, I.; Tomishige, K. Promoting effect of Re addition to $Rh/SiO_2$ on glycerol hydrogenolysis. *Chem. Lett.* 2009, *38*, 540–541

122. Ueda, N.; Nakagawa, Y.; Tomishige, K. Conversion of glycerol to

ethylene glycol over Pt-modified Ni catalyst. *Chem. Lett.* 2010, *39*, 506–507.

123. Tesser, R.; Santacesaria, E.; Di Serio, M.; Di Nuzzi, G.; Fiandra, V. Kinetics of glycerol chlorination with hydrochloric acid: A new route to α,γ-dichlorohydrin. *Ind. Eng. Chem. Res.* 2007, *46*, 6456–6465

124. Lim, J.H.; Song, W.S.; Woo, S.Y.; Lee, D.H. Kinetic model of glycerol chlorination with hydrochloric acid. *Korean J. Chem. Eng.* 2010, *27*, 785–790.

125. Santacesaria, E.; Tesser, R.; Di Serio, M.; Casale, L.; Verde, D. New process for producing epichlorohidrin via glycerol chlorination. *Ind. Eng. Chem. Res.* 2010, *49*, 964–970

126. Lim, J.H.; Song, W.S.; Kwan, M.S.; Woo, S.Y.; Sung, S.W.; Bae, J.W.; Lee, D.H. Modified kinetic model for dichloropropanol synthesis from glycerin and anhydrous HCl at high pressure. *J. Chem. Eng. Jpn.* 2011, *44*, 336–344.

127. Zhao, W.; Yang, B.; Yi, C.; Lei, Z.; Xu, J. Etherification of Glycerol with Isobutylene to Produce Oxygenate Additive Using Sulfonated Peanut Shell Catalyst. *Ind. Eng. Chem. Res.* 2010, *49*, 12399–12404.

128. Vaidya, P.D.; Rodrigues, A.E. Glycerol reforming for hydrogen production: a review. *Chem. Eng. Technol.* 2009, *32*, 1463–1469.

129. Serrano-Ruiz, J.C.; Dumesic, J.A. Catalytic routes for the conversion of biomass into liquid hydrocarbon transportation fuels. *Energy Environ. Sci.* 2011, *4*, 83–99.

# Chapter 4

# Production of Bio-Hydrogenated Diesel by Hydrotreatment of High-Acid-Value Waste Cooking Oil over Ruthenium Catalyst Supported on Al-Polyoxocation-Pillared

Yanyong Liu, Rogelio Sotelo-Boyás, Kazuhisa Murata, Tomoaki Minowa, and Kinya Sakanishi

Biomass Technology Research Center, National Institute of Advanced Industrial Science and Technology, AIST Tsukuba Center 5, 1-1-1 Higashi, Tsukuba, Ibaraki 305-8565, Japan

## ABSTRACT

Waste cooking oil with a high-acid-value (28.7 mg-KOH/g-oil) was converted to bio-hydrogenated diesel by a hydrotreatment process over supported Ru catalysts. Thestandard reaction temperature, $H_2$ pressure, liquid hourly space velocity (LHSV), and $H_2$/oil ratio were 350 °C, 2

MPa, 15.2 h$^{-1}$, and 400 mL/mL, respectively. Both the free fatty acids and the triglycerides in the waste cooking oil were deoxygenated at the same time to form hydrocarbons in the hydrotreatment process. The predominant liquid hydrocarbon products (98.9 wt%) were $n$-$C_{18}H_{38}$, $n$-$C_{17}H_{36}$, $n$-$C_{16}H_{34}$, and $n$-$C_{15}H_{32}$ when a Ru/SiO$_2$ catalyst was used. These long chain normal hydrocarbons had high melting points and gave the liquid hydrocarbon product over Ru/SiO$_2$ a high pour point of 20 °C. Ru/H-Y was not suitable for producing diesel from waste cooking oil because it formed a large amount of $C_5$–$C_{10}$ gasoline-ranged paraffins on the strong acid sites of HY. When Al-polyoxocation-pillared montmorillonite (Al$_{13}$-Mont) was used as a support for the Ru catalyst, the pour point of the liquid hydrocarbon product decreased to −15 °C with the conversion of a significant amount of $C_{15}$–$C_{18}$ $n$-paraffins to *iso*-paraffins and light paraffins on the weak acid sites of Al$_{13}$-Mont. The liquid product over Ru/Al$_{13}$-Mont can be expected to give a green diesel for current diesel engines because its chemical composition and physical properties are similar to those of commercial petro-diesel. A relatively large amount of H$_2$ was consumed in the hydrogenation of unsaturated C=C bonds and the deoxygenation of C=O bonds in the hydrotreatment process. A sulfided Ni-Mo/Al$_{13}$-Mont catalyst also produced bio-hydrogenated diesel by the hydrotreatment process but it showed slow deactivation during the reaction due to loss of sulfur. In contrast, Ru/Al$_{13}$-Mont did not show catalyst deactivation in the hydrotreatment of waste cooking oil after 72 h on-stream because the waste cooking oil was not found to contain sulfur-containing compounds.

# INTRODUCTION

The increase of environmental concerns and the depletion of petroleum reserves have stimulated the search for alternative renewable fuels to avoid climate change and energy shortage [1]. Bio-diesel, which has been recognized as "green fuel", is an important alternative fuel made from renewable resources such as vegetable oil [2]. Although bio-diesel is a renewable fuel with environmental benefits, the utilization of nonfood biomass is important from the viewpoint of food supply [3]. Waste cooking oil is an important biomass resource without competition from food uses [4]. It is presumed that 100–140 kt waste cooking oil from the household sector is discarded every year in Japan [5]. Recently the production of bio-diesel from waste cooking oil has been studied worldwide [6,7,8,9].

Fatty acid methyl ester (FAME) is the first generation bio-diesel and it is produced by the transesterification of vegetable oil with methanol [10]. However, FAME has some shortcomings as a fuel for the current diesel engines because both C=C bonds and C=O bonds remain in the molecules of FAME [11]. The anti-oxidation ability of FAME is low due to unsaturated C=C double bonds, and the flash point of FAME is high as FAME is less flammable than paraffins.

Bio-hydrogenated diesel (abbreviated as BHD) is called the second generation bio-diesel and it is produced by the hydrotreatment of vegetable oil, instead of the transesterification of vegetable oil [12]. BHD is a paraffin mixture because all unsaturated C=C double bonds have been hydrogenated and all oxygen atoms have been eliminated in the hydrotreatment process.

The waste cooking oil usually contains a high concentration of free fatty acids, which greatly increase the cost in the production of FAME type bio-diesel. Base catalysts are highly active in the transesterfication of triglycerides with methanol to produce FAME. However, base catalysts lose their activity in the transesterfication of high-acid-value waste cooking oil as they react with free fatty acids to form soap. Acid-catalyzed transesterification is much slower than base-catalyzed although some new acid catalysts have been developed [6]. In general, a complex two-step process has to be used for the FAME production from high-acid-value waste cooking oil: using acid catalysts to convert free fatty acids in the first step and base catalysts to convert triglycerides in the second step. On the other hand, the conversion of high-acid-value waste cooking oil to hydrocarbon type bio-diesel has an economical advantage in that both free fatty acids and triglycerides are deoxygenated at the same time in a one-step hydrotreatment process [13].

Although vegetable oil can be converted to a mixture of paraffins, cycloparaffins, and aromatic hydrocarbons at high temperatures under high pressures even without a catalyst, a relatively large amount of fatty acids usually remain in the products [14]. Solid acid catalysts (such as H-ZSM-5, $SO_4/ZrO_2$, etc.) can convert vegetable oil to a mixture of gasoline, kerosene, light gas oil, gas oil, and long chain residues by the hydrotreatment process [9,15]. Industrial FCC catalysts can convert vegetable oil to gasoline distilled hydrocarbons in the FCC unit under the FCC conditions [16]. For converting vegetable oil to bio-hydrogenated diesel (BHD), two types of effective catalysts have been reported for the hydrotreatment process: noble metal catalysts (Pd, Pt, etc.) [17,18,19,20,21], and desulfurized catalysts (sulfided Ni-Mo, Co-Mo, Ni-W, Co-W, etc.) [21,22,23,24].

The yield of $C_{15}$–$C_{18}$ n-paraffins usually has been stated in the literature for the hydrotreatment of vegetable oil and it seems that the amount of $C_{15}$–$C_{18}$ n-paraffins determines the quality of BHD fuel [18,19,20,21,22]. Actually, an extreme amount of $C_{15}$–$C_{18}$ n-paraffins give BHD a low fluidity and hinder the use of BHD in current diesel engines. Any post-treatments in adapting BHD for current diesel engines would increase the cost of BHD production. We think that the character of commercial normal diesel from crude oil should be used as a goal for BHD because the current diesel engines have been designed to use normal diesel as a fuel [23]. We have combined sulfided Ni-Mo and solid acids to adjust the composition and properties of BHD in the hydrotreatment of vegetable oils [23,24]. However, the sulfided Ni-Mo catalyst slowly deactivated during the reaction because of sulfur loss [24]. Ru has been reported as an effective noble metal catalyst for desulfurization [25]. Clays are sorts of solid acids with a porous structure and unique acidity [26,27]. Pillaring large metal polyoxocations (such as $[AlO_4Al_{12}(OH)_{24}(H_2O)_{12}]^{7+}$) into the interlayer region of clays (such as montmorillonite) increases the surface area, pore size, and temperature stability of clay materials [27]. Al-polyoxocation-pillared montmorillonite has been reported as an excellent catalyst for producing diesel distilled hydrocarbons from heavy vacuum gas oil and Fischer-Tropsch wax by the hydrotreatment process [28,29].

In this study, we combined Ru with Al-polyoxocation-pillared montmorillonite to convert high-acid-value waste cooking oil to BHD for current diesel engines by the hydrotreatment process, as compared with other supports ($SiO_2$ and H-Y zeolite) and metal catalyst (sulfided Ni-Mo).

# EXPERIMENTAL SECTION

## Materials

Waste vegetable cooking oil was obtained from a restaurant (served fried chicken) and used without any purification. The standard reagents of fatty acids and triglycerides were purchased from Sigma-Aldrich Chemical Company (St. Louis, MO, USA) and Tokyo Kasei Chemical Company (Tokyo, Japan). Chemical reagents for synthesizing catalysts were purchased from Wako Pure Chemical Industries Company (Tokyo, Japan).

Na-type montmorillonite (denoted as Na-Mont) is a natural clay mineral (Kunipia F) supplied by Kunimine Industrial Company (Tokyo, Japan). The cation exchange capacity (CEC) was about 100 meq/100 g and the particle size was smaller than 2 μm in the Na-Mont sample. Na-Y zeolite ($SiO_2/Al_2O_3$ = 5.5; BET surface area: 382 $m^2 \cdot g^{-1}$) was purchased from Wako Pure Chemical Industries Company (Tokyo, Japan). $SiO_2$ support (BET surface area: 300 $m^2 \cdot g^{-1}$; average pore size: 10 nm) were purchased from Fuji Silysia Chemical Company (Tokyo, Japan).

## Catalysts

$[AlO_4Al_{12}(OH)_{24}(H_2O)_{12}]^{7+}$-pillared montmorillonite (denoted as $Al_{13}$-Mont) was prepared by ion-exchange [30,31]. The pillaring solution of $[AlO_4Al_{12}(OH)_{24}(H_2O)_{12}]^{7+}$ was prepared by the dropwise addition of a solution of 0.2 M NaOH (2.4 g NaOH in 300 mL $H_2O$) to a solution of 0.5 M $Al(NO_3)_3$ (9.38 g $Al(NO_3)_3 \cdot 9H_2O$ in 50 mL $H_2O$) until a OH/Al molar ratio of 2.4. After standing at room temperature for 24 h, 5 g Na-Mont was added to the pillaring solution. The final suspension was stirred at room temperature for 12 h and a solid sample was obtained by filtration. The solid sample was washed with distilled water three times, dried at 100 °C for 24 h, and calcined at 400 °C for 3 h.

H-Y was prepared from Na-Y by ion-exchange. Na-Y was stirred in an aqueous solution of $NH_4NO_3$ (0.1 M) for 3 h and $NH_4$-Y was then obtained after filtering from the water. The obtained $NH_4$-Y was dried at 100 °C for 24 h and calcined at 550 °C for 3 h to form H-Y.

Ru-supported catalysts (Ru/SiO$_2$, Ru/H-Y, and Ru/Al$_{13}$-Mont) were synthesized using a wet impregnation method. The calcined solid support (SiO$_2$, H-Y, or Al$_{13}$-Mont) was impregnated with an aqueous solution of RuCl$_3$·$n$H$_2$O (Ru: 40.6 wt%) with stirring. A solid sample was obtained after evaporating the water by heating at 90 °C. The obtained solid sample was dried at 100 °C for 24 h and calcined at 400 °C for 3 h. The designed Ru loading was 1.0 wt% in each catalyst because 5 g solid support and 0.124 g RuCl$_3$·$n$H$_2$O (Ru: 40.6 wt%) in 50 mL water were used for the catalyst preparation. The actual Ru loadings, which were measured by ICP elemental analysis, were similar to the designed Ru loadings in various catalysts. Prior to the reaction, the catalysts were reduced in a H$_2$ flow (flow rate: 60 mL·min$^{-1}$) at 350 °C for 2 h.

Ni-Mo/Al$_{13}$-Mont was synthesized as a reference catalyst using a co-impregnation method [31,32]. Al$_{13}$-Mont support was added to a mixed aqueous solution of (NH$_4$)$_6$Mo$_7$O$_{24}$·4H$_2$O and Ni(NO$_3$)$_2$·6H$_2$O to form a slurry. The slurry was stirred at 90 °C until formation of a solid sample by evaporating water. The obtained solid sample was dried at 100 °C for 24 h and calcined at 400 °C for 3 h. The loadings of MoO$_3$ and NiO were 15 wt% and 3 wt% in Ni-Mo/Al$_{13}$-Mont, respectively. Prior to the reaction, the catalyst was sulfided and reduced in a mixed gas containing 10% H$_2$S and 90% H$_2$ (flow rate: 60 mL·min$^{-1}$) at 400 °C for 10 h.

## Characterization

The elemental composition of waste cooking oil was determined using an elemental analyzer (CE instrument EA1110). The chemical composition of waste cooking oil was analyzed using an Agilent 6890 N FID-GC with an Omnistar Q-mass. A HP-624 capillary column was used to separate free fatty acids and an UA-TRG capillary column was used to separate triglycerides, diglycerides, and monoglycerides. The density of waste cooking oil was determined at 20 °C using a density/specific gravity meter (Kyoto Electronics DA-130N). The viscosity of waste cooking oil was determined at 40 °C using a vibro viscometer (A&D Co. Lim., Tokyo, Japan, SV-10). The acid value of waste cooking oil was determined by an acid-base titration technique using a KOH aqueous solution (ASTM D 664). The limit of detection was 0.1 mg-KOH/g-liquid in the measurement of the acid value. Iodine values of the oils were measured by a titration technique using ICl and Na$_2$S$_2$O$_3$

solutions (ASTM D 1959). Transmission electron microscopy (TEM) images were obtained using a JEOL JEM 2010FX electron microscope equipped with a Hitachi/Keves H-8100/Delta IV EDS operating at 200 kV. The *ex situ* treated samples were supported on Mo grids for the observations. The powder X-ray diffraction (XRD) patterns were measured using a MAC Science MXP-18 diffractometer with Cu Kα radiation operating at 40 kV and 50 mA. Temperature-programmed desorption of ammonia ($NH_3$-TPD) was carried out using a BELCAT-B automatic system equipped with a TCD and an Omnistar Q-mass. A part of the 0.05 g sample was pretreated at 400 °C for 1 h in a He flow with a flow rate of 50 mL·min$^{-1}$. After the temperature was decreased to 100 °C, ammonia was adsorbed onto the sample's surface, followed by evacuation for 1 h at 100 °C to eliminate the weakly adsorbed ammonia. Then, $NH_3$-TPD was recorded from 100 to 600 °C with a rate of 8 °C·min$^{-1}$.

# Reaction

A high-pressure continuous flow fixed-bed reactor was used for the experimental investigation. Figure 1 shows the reaction system used in this study. A stainless steel tubular reactor (*i.d.*: 1 cm; length: 50 cm) was used for loading the catalyst and a furnace was used for heating the tubular reactor. The waste cooking oil was forced into the reactor at a constant rate by a JP-H type high-pressured micro-feeder (Furue Science Company). Meanwhile, a mixed gas containing 90% $H_2$ and 10% Ar (using as an internal standard) was introduced into the reactor from a high-pressure $H_2$ cylinder and the flow rate was controlled by a mass flow controller. A cooling trap (put in ice-water) was set between the reactor exit and the back-pressure regulator to collect liquid products.

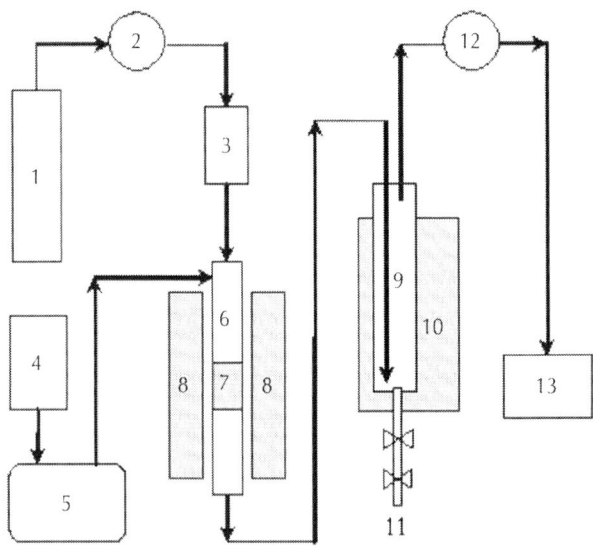

**Figure 1:** Reaction system used in this study. (**1**) H$_2$ cylinder; (**2**) pressure regulator; (**3**) mass flow controller; (**4**) tank of waste cooking oil; (**5**) high-pressured micro-feeder; (**6**) stainless steel tubular reactor; (**7**) catalyst layer; (**8**) furnace; (**9**) cold trap; (**10**) tank of ice-water; (**11**) outlet of liquid product; (**12**) back-pressure regulator; (**13**) on-line GCs.

The catalyst was loaded onto an iron plate with many small holes in the center of the tubular reactor. The pressure in the reaction system was controlled by a back-pressure regulator. The reaction temperature was controlled by a temperature-controller equipped with a thermocouple. The top of the thermocouple was in contact with the surface of the catalyst layer to determine the reaction temperature. The standard reaction conditions were listed as follows: catalyst amount: 1 g (mixed with quartz sand to 5 mL in total volume); reaction temperature: 350 °C; H$_2$ pressure: 2 MPa; LHSV (liquid hourly space velocity, which is a ratio of the volume of liquid feed to the volume of catalyst and quartz sand): 15.2 h$^{-1}$; ratio of H$_2$ to oil in feed: 400 mL/mL, in which the H$_2$ volume was described in the conditions of standard temperature and pressure (STP).

# Analyses

The gas products were continuously analyzed by an on-line GC system during the reaction. Inorganic gases ($H_2$, Ar, CO and $CO_2$) were analyzed using a Shimadzu 8A TCD-GC equipped with MS-5A and Porapak-Q columns. Gaseous hydrocarbons ($C_1$–$C_4$) were analyzed using an Agilent 6890 N FID-GC equipped with a RT-QPLOT capillary column. The factors of various gases were obtained using a standard mixed gas (with known concentration for each component) from a cylinder.

The liquid products were taken out from the cooling trap after the reaction. After removing the water layer in the bottom using a separation funnel, a certain amount of 1-methylnaphthalene was added to the oil phase as an internal standard. The oil phase was then analyzed by an Agilent 6890 N FID-GC equipped with three capillary columns: a UA-DX30 capillary column for analyzing $C_{5+}$ hydrocarbons, a HP-624 capillary column for analyzing fatty acids, and a UA-TRG capillary column for analyzing triglycerides, diglycerides, and monoglycerides.

The yields of $C_1$–$C_4$ hydrocarbons, CO and $CO_2$ in the gas products were calculated using Ar as internal standard. The yields of organic compounds in the liquid products were calculated using 1-methylnaphthalene as internal standard. The yield of water was calculated from the weights of formed water and introduced oil. The carbon mass balance had an error <5% over each catalyst.

# RESULTS AND DISCUSSION

## Characterization of Catalysts

Table 1 summarizes the basic characteristics of the Na-Mont and $Al_{13}$-Mont before and after calcination. The value of the basal plane reflection ($d_{001}$) at the lowest angle in the XRD pattern includes the thickness of a host layer and the gallery height of an interlayer region [26,33]. Because the interlayer cations support the montmorillonite layers, any alternations of interlayer cations certainly cause changes of the $d_{001}$ basal plane in the XRD pattern. The thickness of the host layer is about 9.3 Å for montmorillonite compounds [26,27]. As shown in Table 1, the $d_{001}$ basal plane at 25 °C in the XRD pattern increased from 12.4 Å to 19.2 Å after $Al_{13}$ polyoxocation pillaring. After subtracting the thickness of the host layer (9.3 Å) from the $d_{001}$ spacing, Na-Mont had a gallery height of 3.1 Å and $Al_{13}$-Mont had a gallery height of 9.9 Å. The $Al_{13}$ polyoxocations were successfully introduced into the interlayer region in $Al_{13}$-Mont because the gallery height coincided with the size of $Al_{13}$ polyoxocation [26,27]. The $d_{001}$ value of Na-Mont decreased to 10.6 Å after calcination at 300 °C for 3 h and could not be observed after calcination at 500 °C for 3 h.

**Table 1:** Basic characteristics of raw montmorillonite and its $Al_{13}$-pillared derivative

| Sample | $T_{calcination}$ (°C) | $d_{001}$ (Å) | BET (m²/g) | Total $V_p$ (cm³/g) | Micro$V_p$ (cm³/g) |
|---|---|---|---|---|---|
| Na-Mont | 25 | 12.4 | 82 | 0.079 | 0.004 |
| | 300 | 10.6 | 38 | 0.068 | 0.005 |
| | 400 | 10.5 | 29 | 0.065 | 0.004 |
| | 500 | – | 26 | 0.063 | 0.003 |
| $Al_{13}$-Mont | 25 | 19.2 | 156 | 0.150 | 0.078 |
| | 300 | 18.5 | 237 | 0.170 | 0.111 |
| | 400 | 18.5 | 270 | 0.180 | 0.122 |
| | 500 | 17.9 | 189 | 0.152 | 0.103 |

As shown in Table 1, the BET surface area of $Al_{13}$-Mont was significantly higher than that of its precursor Na-Mont at 25 °C. The height between two clay layers in Na-Mont was only 3.1 Å (from XRD results) and many water molecules existed in the interlayer region at 25 °C. Hence, the interior of the interlayer region in Na-Mont was difficult to utilize for adsorbing $N_2$ molecules in the BET measurement. On the other hand, the gallery height of $Al_{13}$-Mont increased to 9.9 Å (from XRD results) and thus $N_2$ molecules easily enter the interlayer region, which caused a significant increase of BET surface area of $Al_{13}$-Mont at 25 °C. Moreover, the BET surface area of Na-Mont decreased to 38 m²/g after calcination at 300 °C for 3 h due to the collapse of the layered structure. On the other hand, the $d_{001}$ value of $Al_{13}$-Mont slightly decreased to 18.5 Å upon calcination at 300 °C due to the dehydration of $Al_{13}$ polyoxocation. The BET surface area of $Al_{13}$-Mont increased with increasing calcination temperature up to 400 °C, followed by a decrease of surface area upon calcination at 500 °C. Hence, $Al_{13}$-Mont retained the layered structures after calcination at 400 °C owing to the robust polyoxocation pillars but were partly destroyed upon calcination at 500 °C. As for the pore size of a montmorillonite compound, it is defined not only by the interlayer distance (between two clay layers) but also by the lateral distance (between two interlayer pillars). The micropore (pores diameter <20 Å) volume of Na-Mont was very low (about 0.004 cm³/g). On the other hand, $Al_{13}$-Mont is a misroporous material because its micropore volume is relatively large. Both the total pore volume and the micropore volume of $Al_{13}$-Mont increased with increasing calcination temperature up to 400 °C and decreased upon calcination at 500 °C due to partial destruction of the layered structure. Therefore, the calcination temperature for $Al_{13}$-Mont-containing catalysts was set at 400 °C in this study.

Figure 2 shows the TEM image of Ru/$Al_{13}$-Mont (Ru loading: 1 wt%) catalyst after reduction in a $H_2$ flow at 300 °C for 2 h. The layered structure of montmorillonite support could be observed from the TEM image. The gallery height was about 1 nm and this value coincided with the size of $Al_{13}$ polyoxocation. The black Ru particles could be observed on the surface of the montmorillonite support. The Ru particles did not enter into the interlayer region of montmorillonite in Ru/$Al_{13}$-Mont. The distribution of Ru particles was relatively uniform and the particle size of Ru was about 3–5 nm.

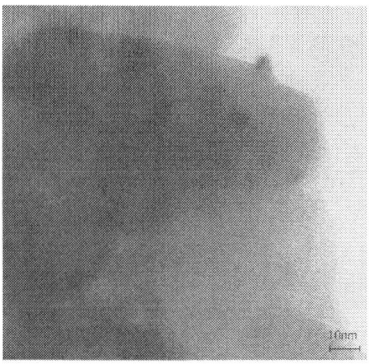

**Figure 2:** TEM image of Ru/$Al_{13}$-Mont (Ru loading: 1 wt%) catalyst after reduction.

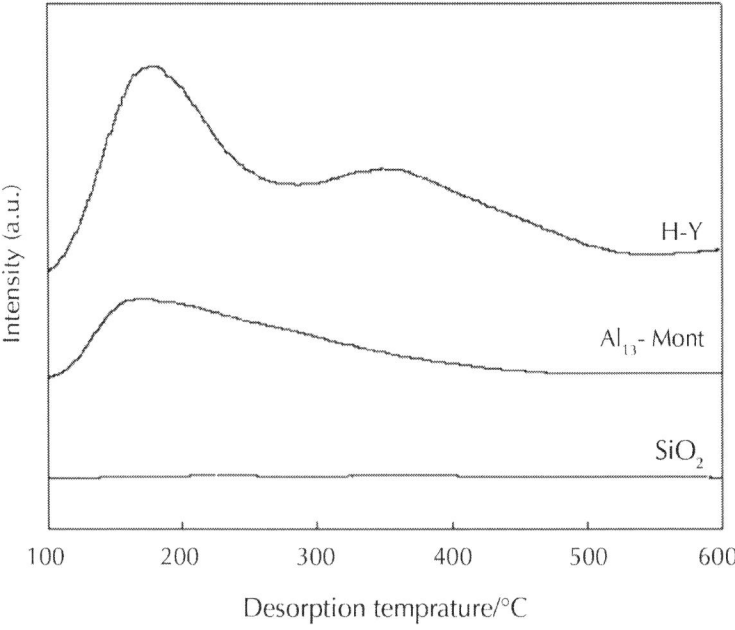

**Figure 3:** NH$_3$-TPD profiles of various solid supports.

Figure 3 shows the NH$_3$-TPD of various solid supports used in this study. NH$_3$-TPD is a powerful tool for estimating the acidic property of a solid surface. The adsorbed NH$_3$ molecules desorbed from weak acid sites at low temperatures and desorbed from strong acid sites at high temperatures. As shown in Figure 3, SiO$_2$ did not show an obvious peak in the NH$_3$-TPD profile, indicating that SiO$_2$ did not have acid sites on the surface. Al$_{13}$-Mont had acid sites on the surface area because its NH$_3$-TPD showed a peak with a maximum value at 170 °C. Because H-Y showed two peaks at about 180 and 350 °C, H-Y had two types of acid sites (weak acid sites and strong acid sites) on the surfaces. The peak at the maximum temperature in the NH$_3$-TPD profile corresponds to the strongest acid sites on the solid surface. The nature of a solid acid catalyst is mainly determined by the strongest acid sites on the surface. According to the peak position at the maximum temperature in the NH$_3$-TPD profiles, the acidic strength of various supports was in an order of H-Y > Al$_{13}$-Mont > SiO$_2$ (no acidity).

# Composition and Property of Waste Cooking Oil

Table 2 shows the composition and property of the waste cooking oil used in this study.

**Table 2:** Composition and property of the waste cooking oil used in this study

| Elemental composition | C | H | N | S | O |
|---|---|---|---|---|---|
| Weight percent (wt%) | 77.91 | 11.69 | 0.04 | 0 | 10.36 |
| Chemical composition | Triglyceride | Diglyceride | Monoglycerid | Free fatty acid | |
| Content (g/100 g-oil) | 79.1 | 1.8 | 2.2 | 16.9 | |
| Others | Acid value (mg-KOH/g-oil) | Iodine value (g-I₂/100 g-oil) | Viscosity (mPa/s) | Density (g/mL) | |
| | 28.7 | 88.6 | 57.8 | 0.92 | |

The elemental composition of the waste cooking oil was determined using an elemental analyzer (CE instrument EA1110). The difference from 100% was taken as the content of oxygen. As shown in Table 2, the amount of N was very small (0.04 wt%) and S could not be observed in the waste cooking oil. In general, vegetable oils do not contain sulfur-containing compounds. Therefore, the bio-diesels produced from vegetable oils are regarded as environmentally benign green fuels. Concurrently, the development of catalysts for the hydrotreatment of vegetable oils (without S) is different from those for the hydrotreatment of fossil fuels with sulfur-containing compounds.

The chemical components in the waste cooking oil were confirmed by GC-MS results and standard reagents. The content of free fatty acids was very high (16.9 g/100 g-oil) in the waste cooking oil. It seems that the waste cooking oil can not be used as a straight vegetable oil (SVO) just after filtration because the free fatty acids are corrosive to diesel engines. For the FAME production, base catalysts lose their activity in the transesterfication of cooking vegetable oil because of soap formation from free fatty acids and basic catalysts. A complex two-step process has to be used for the FAME production (using acid catalyst to convert free fatty acids in the first step and using base catalyst to convert triglycerides in the second step) which must increase the cost of the FAME production. Therefore, the one-step hydrotreatment process is suitable for the commercial production of BHD from waste cooking oil. As for the physical properties, the acid value of the waste cooking oil was very high (28.7 mg-KOH/g-oil) because it contained a large amount of free fatty acids. The iodine value of the waste cooking oil was 88.6 g-$I_2$/100 g-oil, indicating that the waste cooking oil contained many C=C unsaturated bonds. The viscosity at 30 °C was 57.8 mPa/s and the density at 25 °C was 0.92 g/mL for the waste cooking oil. The viscosity was too high to use the waste cooking oil in current diesel engines.

Table 3 shows the species of free fatty acid (FFA) in the waste cooking oil used in this study. The formula of each fatty acid was confirmed by GC-MS (HP-624 capillary column) results with standard reagents. The content of each fatty acid shown in Table 3 was expressed as its weight in 100 g free fatty acid, while 100 g waste cooking oil contained 16.9 g free fatty acid (Table 2). The main free fatty acids in the waste cooking oil were palmitic acid (6.6%), stearic acid (7.8%), oleic acid (53.6%), linoleic acid (16.4%), and linolenic acid (14.3%). The unsaturated fatty acids (with C=C double bond) included palmitoleic, oleic, linoleic, linolenic, and eicosenoic acids, while the saturated fatty acids (without C=C double bond) included palmitic and stearic acids. By calculating using these data, the unsaturated fatty acids occupied 85.6% and the saturated fatty acids occupied 14.4% in total free fatty acids of the waste cooking oil. All free fatty acids in the waste cooking oil had even carbon numbers (16, 18, and 20) in the carbon chains. The fatty acids with sixteen carbons in the carbon chain (so-called $C_{16}$-acids, including palmitic and palmitoleic acids) occupied 6.8%, the fatty acids with eighteen carbons in the carbon chain (so-called $C_{18}$-acids, including stearic, oleic, linoleic and linolenic acids) occupied 92.1%, and the fatty acids with twenty carbons in the carbon chain (so-called $C_{20}$-acids, including eicosenoic acid) occupied 1.1% in total free fatty acids of the waste cooking oil.

**Table 3:** Species of free fatty acid (FFA) in the waste cooking oil used in this study

| Formula | Name | Structure [a] | Content (g/100 g-FFA) |
|---------|------|-----------|------------------------|
| $C_{16}H_{32}O_2$ | Palmitic | C16:0 | 6.6 |
| $C_{16}H_{30}O_2$ | Palmitoleic | C16:1 | 0.2 |
| $C_{18}H_{36}O_2$ | Stearic | C18:0 | 7.8 |
| $C_{18}H_{34}O_2$ | Oleic | C18:1 | 53.6 |
| $C_{18}H_{32}O_2$ | Linoleic | C18:2 | 16.4 |
| $C_{18}H_{30}O_2$ | Linolenic | C18:3 | 14.3 |
| $C_{20}H_{38}O_2$ | Eicosenoic | C20:1 | 1.1 |

# Hydrotreatment of Waste Cooking Oil over Ru-Supported Catalysts

Table 4 shows the product yields over various catalysts in the hydrotreatment of waste cooking oil. The conversion was 100% and did not decrease after reaction for 10 h over each catalyst because triglycerides and fatty acid acids could not be detected in the products. The yield of fuel gas ($C_1$ + $C_2$) was very low (<1.0 wt%) over each catalyst. The yield of liquid fuel ($C_{5+}$ hydrocarbons) was in a narrow range of 82.1–84.0 wt% and the yield of LPG ($C_3$ + $C_4$) was in a narrow range of 4.8–5.6 wt% over each catalyst. Both liquid hydrocarbons ($C_{5+}$) and LPG ($C_3$ + $C_4$) can be used as fuels for automobiles. Water (yield: 7.7–8.0 wt%) and $CO_x$ (including CO and $CO_2$, yield: 3.2–3.4 wt%) were also formed in the hydrotreatment process over each catalyst. Because propane is formed after all C=O bonds in triglycerides are broken during the hydrotreatment process, propane occupied more than 90 wt% in the formed LPG ($C_3$ + $C_4$) over each catalyst.

**Table 4:** Product yields over various catalysts in the hydrotreatment of waste cooking oil

| Catalyst | $C_1$ + $C_2$ | $C_3$ + $C_4$ | $C_{5+}$ | $CO_x$ | $H_2O$ |
|----------|---------------|---------------|----------|--------|--------|
| Ru/SiO$_2$ | 0.2 | 4.8 | 84.0 | 3.3 | 7.7 |

| Ru/Al$_{13}$-Mont | 0.4 | 5.0 | 83.6 | 3.2 | 7.8 |
|---|---|---|---|---|---|
| Ru/H-Y | 0.9 | 5.6 | 82.1 | 3.4 | 8.0 |

The waste cooking oil contained triglycerides, diglycerides, monoglycerides, and free fatty acids (Table 2). The hydrotreatment process on the Ru sites contained two steps: hydrogenation and deoxygenation. In the first step, all unsaturated C=C bonds in the triglycerides, diglycerides, monoglycerides, and free fatty acids were saturated on their carbon chains by the hydrogenation on the Ru sites. In the second step, the saturated triglycerides decomposed by scission of the C=O bonds, leading to the formation of diglycerides, monoglycerides, and carboxylic acids (fatty acids) in order, and the carboxylic acids then underwent deoxygenation on the Ru sites to form hydrocarbons. Therefore, the key step in the hydrotreatment of waste cooking oil is the deoxygenation of saturated fatty acids.

The deoxygenation of saturated fatty acids (such as stearic acid $C_{17}H_{35}COOH$) contains three parallel reactions: reduction, decarbonylation, and decarboxylation [22].

| $C_{17}H_{35}COOH + 3H_2 = C_{18}H_{38} + 2H_2O$ | (Reduction) | (Reaction 1) |
|---|---|---|
| $C_{17}H_{35}COOH + H_2 = C_{17}H_{36} + CO + H_2O$ | (Decarbonylation) | (Reaction 2) |
| $C_{17}H_{35}COOH = C_{17}H_{36} + CO_2$ | (Decarboxylation) | (Reaction 3) |

As shown in Reactions 1–3, for a fatty acid with an even carbon number, the reduction produces a normal paraffin with an even carbon number plus water; the decarbonylation produces a normal paraffin with an odd carbon number plus water and CO; and the decarboxylation produces a normal paraffin with an odd carbon number plus $CO_2$. Both the decarbonylation and the decarboxylation occurred during the reaction because both CO and $CO_2$ were detected in the gas product over each catalyst.

As shown in Table 4, the three catalysts had similar yields of $C_{5+}$ liquid hydrocarbons (ranging from 82.1 to 84.0 wt%) from the hydrotreatment of waste cooking oil. However, it is necessary to investigate whether these liquid hydrocarbons are suitable for use as a diesel fuel in current diesel engines.

Figure 4 shows the FID-GC charts (UA-DX capillary column) of liquid hydrocarbons formed from the hydrotreatment of waste cooking oil over various catalysts. Ru/SiO$_2$ formed $n$-C$_{18}$H$_{38}$, $n$-C$_{17}$H$_{36}$, $n$-C$_{16}$H$_{34}$, and $n$-C$_{15}$H$_{32}$ as the main products and the amounts of $iso$-paraffins and light paraffins ($\leq$C$_{14}$) were very low. According to Reactions 1–3, $n$-C$_{16}$H$_{34}$ and $n$-C$_{18}$H$_{38}$ were formed by the reduction of palmitic acid and stearic acid, and $n$-C$_{15}$H$_{32}$ and $n$-C$_{17}$H$_{36}$ were formed by the decarbonylation and decarboxylation of palmitic acid and stearic acid, respectively. In order to adjust the chemical composition of the liquid hydrocarbon product, we supported Ru catalyst on solid acids (H-Y, Al$_{13}$-Mont) for the isomerization/cracking of C$_{15}$H$_{32}$–C$_{18}$H$_{38}$ $n$-paraffins as the third step after the hydrogenation and deoxygenation steps. The isomerization/cracking of $n$-paraffins over bifunctional catalysts containing metal and solid acid is very important for improving the properties of liquid hydrocarbon fuels [34,35,36,37]. The reaction proceeds via carbenium ion intermediates formed on the solid acid sites. The acidic strength of the solid acid determines the activity of the bifunctional catalysts when the same metal is used in the isomerization/cracking of $n$-paraffins. Because the acidic strength of various supports was in the order of H-Y > Al$_{13}$-Mont > SiO$_2$ (Figure 3), Ru/H-Y formed a large amount of light hydrocarbons ($\leq$C$_{14}$) and Ru/Al$_{13}$-Mont formed a significant amount of light hydrocarbons ($\leq$C$_{14}$) (Figure 4).

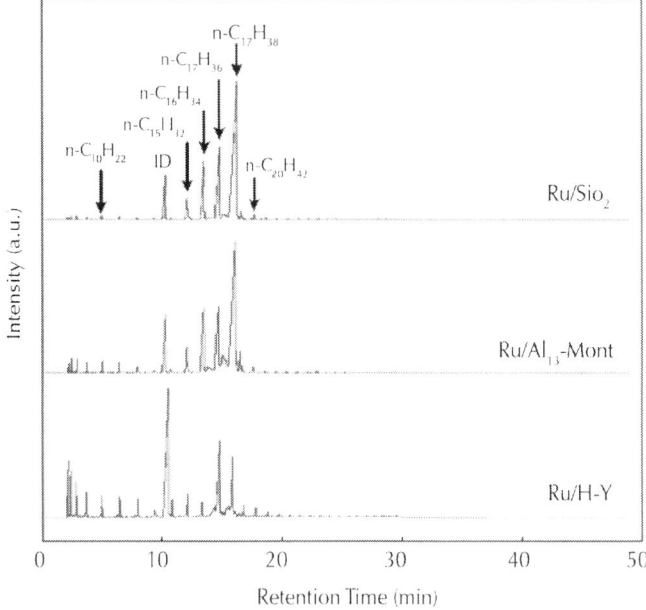

**Figure 4:** FID-GC charts of liquid hydrocarbons formed from the hydrotreatment of waste cooking oil over various catalysts (reaction conditions: same as those in Table 4).

**Table 5:** Composition and property of the liquid hydrocarbons ($C_{5+}$) formed from the hydrotreatment of waste cooking oil over various catalysts [a]

| Catalyst | Composition (wt%) | | | Iso/n ratio | Pour Point (°C) | Property | |
|---|---|---|---|---|---|---|---|
| | $C_5$–$C_{10}$ | $C_{11}$–$C_{20}$ | $C_{20+}$ | | | Density at 25 °C (g/mL) | Viscosity at 30 °C (mPa/s) |
| Ru/SiO$_2$ | 0.9 | 98.9 | 0.2 | 0.08 | 20 | 0.79 | 8.01 |
| Ru/Al$_{13}$-Mont | 9.1 | 89.8 | 1.1 | 0.22 | −15 | 0.78 | 3.96 |
| Ru/H-Y | 42.8 | 56.5 | 0.7 | 0.43 | —[b] | 0.77 | 2.08 |
| Normal diesel | 8.2 | 88.1 | 3.7 | 0.28 | −15 | 0.82 | 3.69 |

[c]

Table 5 shows the composition and property of the liquid hydrocarbons ($C_{5+}$) formed from the hydrotreatment of waste cooking oil over various catalysts. Although the liquid hydrocarbons formed over $Ru/SiO_2$ contained the largest amount of $C_{11}$–$C_{20}$ diesel-distillate (98.9 wt%) with the lowest *iso/n* ratio (0.08) among various catalysts, the pour point of the product was too high (20 °C) to use as a diesel fuel. Ru/H-Y was also not suitable for producing BHD from the waste cooking oil because it formed a large amount of $C_5$–$C_{10}$ gasoline-distillate (42.8 wt%) on the strong acid sites. $Ru/Al_{13}$-Mont produced a liquid hydrocarbon product with a pour point of −15 °C with an *iso/n* ratio of 0.22.

The melting points of $n$-$C_{18}H_{38}$, $n$-$C_{17}H_{36}$, $n$-$C_{16}H_{34}$, and $n$-$C_{15}H_{32}$ are 28, 22, 18, and 10 °C, respectively. A predominant amount of $C_{15}$–$C_{18}$ $n$-paraffins gave a high pour point (20 °C) for the liquid hydrocarbon product over $Ru$-$Mo/SiO_2$ (Table 5). It is necessary to decrease the pour point of the product over $Ru/SiO_2$ in order to use it in the current diesel engines. *Iso*-paraffins and light paraffins have relatively low melting points (for example, 2-methyl-heptadecane: 5 °C; 3-methyl-heptadecane: −6 °C; 2-methyl-hexadecane: 5 °C; 2-methyl-pentadecane: −11 °C; 3-methyl-pentadecane: −22 °C; 2-methyl-tetradecane: −8 °C; 3-methyl-tetradecane: −36 °C; $n$-$C_{11}H_{24}$: −25 °C). Hence, we supported Ru on solid acids ($Al_{13}$-Mont and H-Y) to improve the fluidity of the liquid hydrocarbon product from the hydrotreatment of waste oil by the isomerization/cracking of $C_{15}H_{32}$–$C_{18}H_{38}$ $n$-paraffins.

The hydrocarbons ranging from $C_{11}$ to $C_{20}$ are called diesel-distillate and the hydrocarbons ranging from $C_5$ to $C_{10}$ are called gasoline-distillate. However, the commercial normal diesel bought from a petrol station actually contains 8.2 wt% of $C_5$–$C_{10}$ gasoline-distillate to improve the fluidity (Table 5). The viscosity of the liquid hydrocarbon product over $Ru/Al_{13}$-Mont was 3.96 mPa/s, which was quite similar to that of the normal diesel (3.69 mPa/s). The density of the normal diesel was relatively large (0.82 g/mL), probably because the normal diesel contained some heavy additives. On the whole, both composition and pour point of the liquid hydrocarbon product over $Ru/Al_{13}$-Mont were quite similar to those of the normal diesel bought from a petrol station. Because the current diesel engines have been designed to use normal diesel as a fuel, we think that the product over $Ru/Al_{13}$-Mont (with similar composition and property to normal diesel) is better than the product over Ru/SiO$_2$ (with 98.9 wt% C$_{11}$–C$_{20}$ diesel-distillate and high pour point) as a diesel fuel for current diesel engines. As a result, because Al$_{13}$-Mont had a significant acidic strength, the liquid hydrocarbon product in the hydrotreatment of waste cooking oil over Ru/Al$_{13}$-Mont can be expected to give a green BHD fuel directly usable in the current diesel engines without any post-treatment processes (such as distillation, upgrading, etc.).

# Influence of Reaction Conditions in the Hydrotreatment of Waste Cooking Oil over Ru/Al$_{13}$-Mont

Because the molecules of triglycerides and free fatty acids in the waste cooking oil contained many unsaturated C=C bonds and C=O bonds, the H$_2$ amount and H$_2$ pressure are important reaction conditions for the hydrotreatment of waste cooking oil to produce saturated hydrocarbons.

Figure 5 shows the effect of H$_2$/oil ratio in the hydrotreatment of waste cooking oil over Ru/Al$_{13}$-Mont. The H$_2$/oil was a ratio of H$_2$ feed rate to liquid (waste cooking oil) feed rate during the reaction and the H$_2$ volume was described in conditions of standard temperature and pressure (STP). A low H$_2$/oil is desirable for producing BHD for commercialization as it is favorable to reduce the cost of production. However, the peaks of fatty acids could be detected in the GC chart at a H$_2$/oil ratio of 300 (Figure 5). Thus a H$_2$/oil ratio of 400 was necessary for the hydrotreatment of waste cooking oil over Ru/Al$_{13}$-Mont. As shown in Reactions 1–3, the deoxygenation of fatty acids contained three parallel reactions: reduction, decarbonylation and decarboxylation. The reduction does not produce CO$_x$, while both the decarbonylation and the decarboxylation produces CO$_x$ and thus one C in the carbon chain is lost. Thus from the deoxygenation of C$_{16}$-acids and C$_{18}$-acids, the reduction produces $n$-C$_{16}$H$_{34}$ and $n$-C$_{18}$H$_{38}$, and the decarbonylation and decarboxylation produce $n$-C$_{15}$H$_{32}$ and $n$-C$_{17}$H$_{36}$. Moreover, the reduction is favorable under a large H$_2$/oil ratio because it consumes more H$_2$ molecules than the decarbonylation and decarboxylation (Reactions 1–3). Hence, the ratios of C$_{18}$/C$_{17}$ and C$_{16}$/C$_{15}$ in products over Ru/Al$_{13}$-Mont decreased with decreasing H$_2$/oil ratio from 500 to 300 (Figure 5).

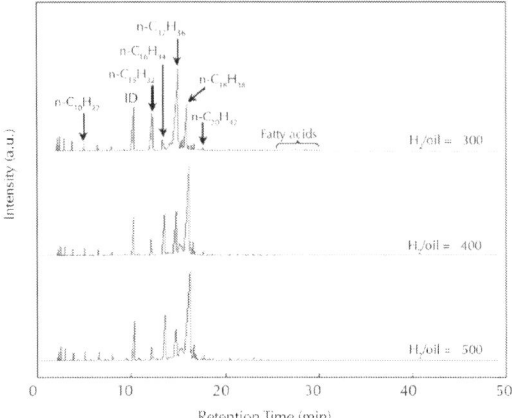

**Figure 5:** FID-GC charts of liquid products in the hydrotreatment of waste cooking oil over Ru/Al$_{13}$-Mont under various H$_2$/oil ratios (*T*: 350 °C; H$_2$ pressure: 2 MPa; liquid hourly space velocity (LHSV): 15.2 h$^{-1}$).

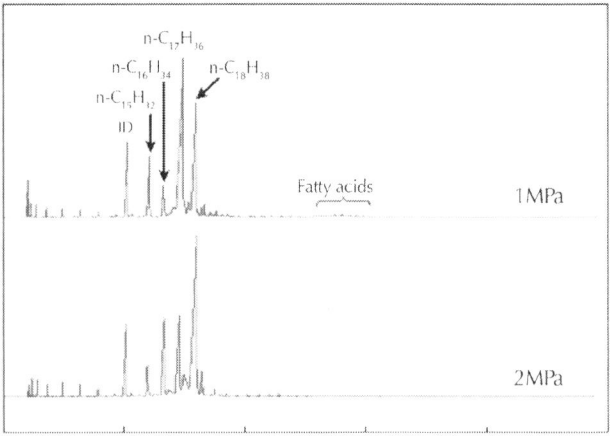

**Figure 6:** FID-GC charts of liquid products in the hydrotreatment of waste cooking oil over Ru/Al$_{13}$-Mont under various H$_2$ pressures (*T*: 350 °C; H$_2$/oil: 400; LHSV: 15.2 h$^{-1}$).

Figure 6 shows the effect of $H_2$ pressure in the hydrotreatment of waste cooking oil over Ru/Al$_{13}$-Mont. A low $H_2$ pressure is desirable for producing BHD in commercialization because it is favorable to reduce the investment in plant and equipment. The peaks of fatty acids could not be observed in the GC chart under 2 MPa of $H_2$ pressure but could be detected in the GC chart under 1 MPa of $H_2$ pressure. Hence, a $H_2$ pressure of 2 MPa was necessary for the hydrotreatment of waste cooking oil over Ru/Al$_{13}$-Mont. As shown in Reactions 1–3, the reduction (for producing $n$-C$_{18}$H$_{38}$ and $n$-C$_{16}$H$_{34}$) is favorable under a high $H_2$ pressure in comparison with the decarbonylation and decarboxylation (for producing $n$-C$_{17}$H$_{36}$ and $n$-C$_{15}$H$_{32}$). Thus the ratios of C$_{18}$/C$_{17}$ and C$_{16}$/C$_{15}$ in products over Ru/Al$_{13}$-Mont decreased with decreasing $H_2$ pressure.

Figure 7 shows the effect of reaction temperature in the hydrotreatment of waste cooking oil over Ru/Al$_{13}$-Mont. When the reaction was carried out at a low reaction temperature of 300 °C, the peaks of fatty acids were observed in the GC chart. This means that the deoxygenation of waste cooking oil was not enough at 300 °C. The peaks of fatty acids could not be observed in GC charts at high reaction temperatures of 350 °C and 400 °C. Although $n$-C$_{18}$H$_{38}$, $n$-C$_{17}$H$_{36}$, $n$-C$_{16}$H$_{34}$, and $n$-C$_{15}$H$_{32}$ are the main products over all three catalysts, the amount of light hydrocarbons ($\leq$C$_{14}$) formed remarkably increased when the reaction temperature increased from 300 to 400 °C. We think the cracking of C$_{15}$H$_{32}$–C$_{18}$H$_{38}$ nomal paraffins caused the increase of light hydrocarbons ($\leq$C$_{14}$) at high reaction temperatures. Hence, the reaction temperature also can be used to adjust the composition and property of the BHD product. Because the amount of C$_5$–C$_{10}$ gasoline-distillate hydrocarbons was too much at 400 °C, the most suitable reaction temperature was found to be 350 °C in the hydrotreatment of waste cooking oil.

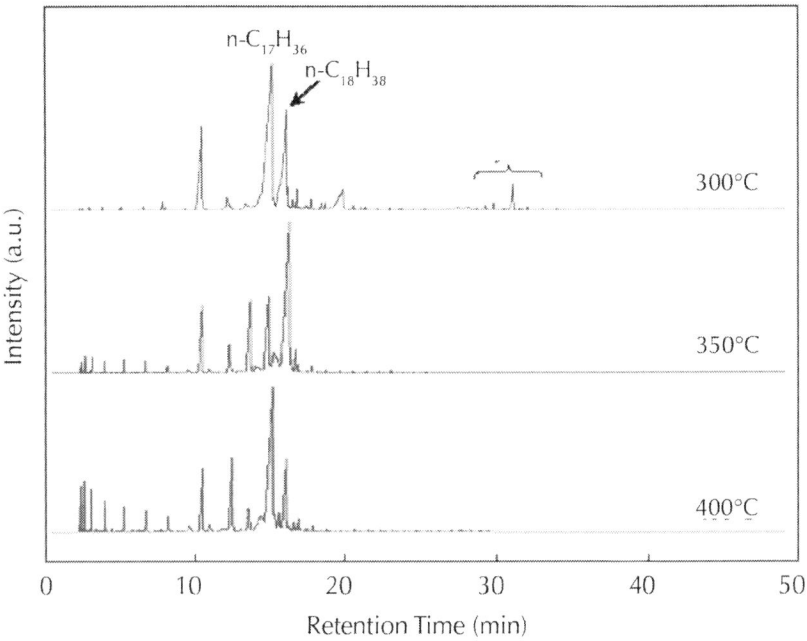

**Figure 7:** FID-GC charts of liquid products in the hydrotreatment of waste cooking oil over Ru/Al$_{13}$-Mont at various reaction temperatures (H$_2$ pressure: 2 MPa; H$_2$/oil: 400; LHSV: 15.2 h$^{-1}$).

The goal of this study is to produce green BHD fuel for use in current diesel engines without any post-treatment. The liquid organic phase in the product should not contain any fatty acids because they are corrosive to engines. Moreover, a significant amount of C$_5$–C$_{10}$ gasoline-distillate (similar to the amount in commercial normal diesel) should be contained in order to adjust the physical properties of BHD fuel to fit current diesel engines. Hence, we chose a H$_2$/oil ratio of 400 mL/mL, a H$_2$ pressure of 2 MPa, and a reaction temperature of 350 °C as the standard reaction conditions in this study.

# Deactivation of Ru/Al$_{13}$-Mont and Sulfided Ni-Mo/Al$_{13}$-Mont Catalysts

Figure 8 shows the time sequences in the hydrotreatment of waste cooking oil over Ru/Al$_{13}$-Mont and sulfided Ni-Mo/Al$_{13}$-Mont. Ru/Al$_{13}$-

Mont showed an initial $C_{5+}$ hydrocarbons yield of 83.6% and it did not decrease after reaction for 72 h. This indicated that Ru/Al$_{13}$-Mont had a high catalytic stability for the hydrotreatment of waste cooking oil. In contrast, sulfided Ni-Mo/Al$_{13}$-Mont showed an initial $C_{5+}$ hydrocarbons yield of 83.5% but it slowly decreased to 76.8% after reaction for 72 h. Ni-Mo/Al$_{13}$-Mont was sulfided by 10% $H_2S$ (in 90% $H_2$) at 400 °C prior to the reaction and a Ni-Mo-S phase was formed as the active phase for the hydrotreatment of waste cooking oil. The sulfur component in sulfided Ni-Mo/Al$_{13}$-Mont was lost slowly during the reaction [24]. Because the waste cooking oil used in this study did not contain a sulfur component (Table 2), the loss of sulfur could not be supplemented by the feedstock. As shown in Figure 8, the speed of deactivation was slow in the first 10 h and became fast after reaction for 10 h over sulfided Ni-Mo/Al$_{13}$-Mont. The loss of sulfur caused the collapse of the Ni-Mo-S active phase in sulfided Ni-Mo/Al$_{13}$-Mont. As for Ru/Al$_{13}$-Mont, it could maintain activity during the reaction because the feedstock did not contain any sulfur poisons.

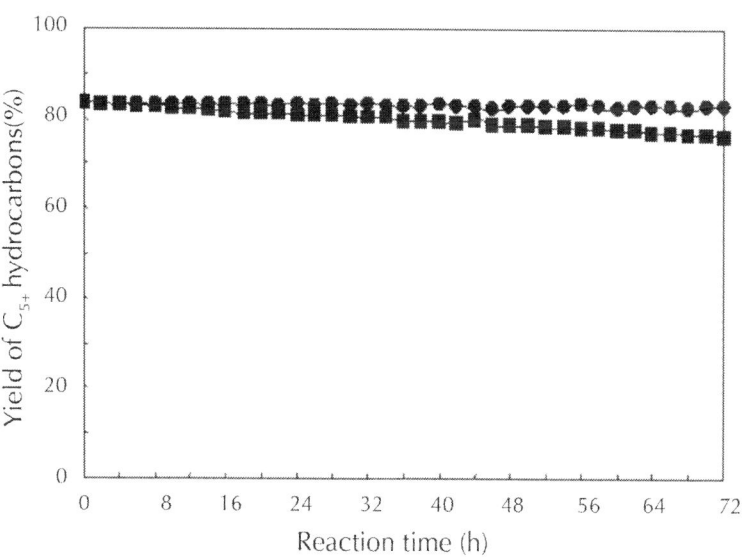

**Figure 8:** Time courses in the hydrotreatment of waste cooking oil over Ru/Al$_{13}$-Mont (●) and sulfided Ni-Mo/Al$_{13}$-Mont (■) (reaction conditions: same as those in Table 4).

Figure 9 shows the time sequences in the hydrotreatment of waste cooking oil containing 10 ppm dimethyl disulfide (DMDS) over Ru/$Al_{13}$-Mont and sulfided Ni-Mo/$Al_{13}$-Mont. A large number of liquid fuels produced from crude oil, coal, and biomass contain sulfur compounds. We added 10 ppm DMDS as a standard sulfur compound in the waste cooking oil to check the influence of sulfur on Ru/$Al_{13}$-Mont and sulfided Ni-Mo/$Al_{13}$-Mont catalysts. As shown in Figure 9, Ru/$Al_{13}$-Mont showed an initial $C_{5+}$ hydrocarbons yield of 83.6% and it decreased to 75.8% after reaction for 72 h for the hydrotreatment of waste cooking oil containing 10 ppm DMDS. Because S rapidly reacted with the Ru atoms to form an inactive RuS phase on the catalyst surface, the speed of deactivation over Ru/$Al_{13}$-Mont was fast in the first 10 h and then became slow. In contrast, sulfided Ni-Mo/$Al_{13}$-Mont showed an initial $C_{5+}$ hydrocarbons yield of 83.5% and it did not decrease after reaction for 72 h. The sulfur compound (DMDS) in the feed waste cooking oil supplemented the loss of sulfur from the surface of sulfided Ni-Mo/$Al_{13}$-Mont catalyst. The structure the Ni-Mo-S active phase remained during the reaction and maintained the catalytic activity of sulfided Ni-Mo/$Al_{13}$-Mont. Some researchers have carried out the hydrotreatment of vegetable oil (not containing S) mixed with heavy vacuum oil (containing S) over Ni-Mo catalysts in order to maintain the catalyst stability [22]. Hence, in the hydrotreatment process, noble catalysts (Ru, Pt, Pd, *etc.*) are suitable for the feedstock without a S component (such as vegetable oils, F-T waxes, *etc.*), and desulfurizated catalysts (sulfided Ni-Mo, Co-Mo, Ni-W, Co-W, *etc.*) are suitable for the feedstock containing S compounds (such as heavy oil, bio oil, *etc.*).

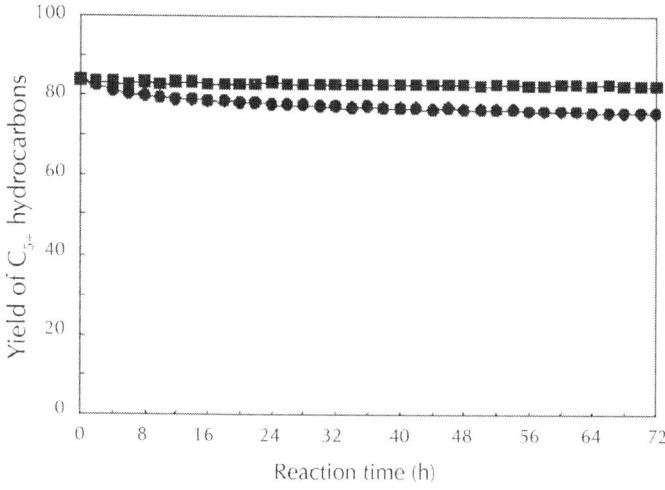

**Figure 9:** Time sequences in the hydrotreatment of waste cooking oil containing 10 ppm DMDS over Ru/Al$_{13}$-Mont (●) and sulfided Ni-Mo/Al$_{13}$-Mont (■) (reaction conditions: same as those in Table 4).

# CONCLUSIONS

The waste cooking oil containing 16.9 wt% free fatty acids was converted to mixed paraffins over Ru-supported catalysts by a one-step hydrotreatment process in which both free fatty acids and triglycerides were deoxygenated at the same time. The fluidity of the liquid hydrocarbon product over Ru/SiO$_2$ was poor because of a predominant amount of C$_{15}$H$_{32}$–C$_{18}$H$_{38}$ n-paraffins. By supporting Ru on acidic Al$_{13}$-Mont, both the chemical composition and the pour point of the liquid hydrocarbon product became quite similar to those of a commercial normal diesel. Ru/Al$_{13}$-Mont acted as a multifunctional catalyst enabling hydrogenation, deoxidization, and isomerization/ cracking for the hydrotreatment of waste cooking oil. The lowest H$_2$/ oil ratio was 400 and the lowest H$_2$ pressure was 2 MPa to convert the waste cooking oil to saturated hydrocarbons over Ru/Al$_{13}$-Mont as the waste cooking oil contained many unsaturated C=C bonds and C=O bonds. Ru/Al$_{13}$-Mont did not deactivate after reaction for 72 h because the waste cooking oil did not contain S compounds as poisons for Ru.

# REFERENCES

1.  Fernando, S.; Adhikari, S.; Chandrapal, C.; Murali, N. Biorefineries: Current status, challenges, and future direction. *Energy Fuel* **2006**, *20*, 1727–1737.

2.  Sharma, Y.C.; Singh, B.; Upadhyay, S.N. Advancements in development and characterization of biodiesel: A review. *Fuel* **2008**, *87*, 2355–2373.

3.  Pinzi, S.; Garcia, L.; Lopez-Gimenez, F.J.; Luque de Castro, M.D.; Dorado, G.; Dorado, M.P. The ideal vegetable oil-based biodiesel composition: A review of social, economical and technical implications. *Energy Fuel* **2009**, *23*, 2325–2341.

4.  Kulkarni, M.G.; Dalai, A.K. Waste cooking oil—An economical source for biodiesel: A review. *Ind. Eng. Chem. Res.* **2006**, *45*, 2901–2913.

5.  Japanese government biomass policies: Major source of introduction of biofuel. *Report on Promotion of Eco-Fuel for Transportation*; Japanese Government: Tokyo, Japan, 2006.

6.  Cao, F.; Chen, Y.; Zhai, F.; Li, J.; Wang, J.; Wang, X.; Wang, S.; Zhu, W. Biodiesel production from high acid value waste frying oil catalyzed by superacid heteropolyacid. *Biotechnol. Bioeng.* **2008**, *101*, 93–100.

7.  Toba, M.; Abe, Y.; Kuramochi, H.; Osako, M.; Mochizuki, T.; Yoshimura, Y. Hydrodeoxygenation of waste vegetable oil over sulfide catalysts. *Catal. Today* **2011**, *164*, 533–537.

8.  Xu, J.; Xiao, G.; Zhou, Y.; Jiang, J. Production of biofuels from high-acid-value waste oils. *Energy Fuel* **2011**, *25*, 4638–4642.

9.  Charusiri, W.; Vitidsant, T. Kinetic study of used vegetable oil to liquid fuels over sulfated zirconia. *Energy Fuel* **2005**, *19*, 1783–1789.

10. Ma, F.; Hanna, M.M. Biodiesel production: A review. *Bioresour. Technol.* **1999**, *70*, 1–15.

11. Sharma, Y.C.; Singh, B.; Upadhyay, S.N. Advancements in development and characterization of biodiesel: A review. *Fuel* **2008**, *87*, 2355–2373.

12. Kalnes, T.; Markery, T.; Shonnard, D.R. Green diesel: A second

generation biofuel. *Int. J. Chem. React. Eng.* **2007**, *5*, A48.

13. Guzman, A.; Torres, J.E.; Prada, L.P.; Nunez, M.L. Hydropressing of crude palm oil at pilot plant scale. *Catal. Today* **2010**, *156*, 38–43.]

14. Li, L.; Coppola, E.; Rine, J.; Miller, J.L.; Walker, D. Catalytic hydrothermal conversion of triglycerides to non-ester biofuels. *Energy Fuel* **2010**, *24*, 1305–1315.

15. Charusiri, W.; Vitidsant, T. Catalytic cracking of used cooking oil to liquid fuels over HZSM-5. *J. Energy* **2003**, *5*, 58–68.

16. Melero, J.A.; Clavero, M.M.; Calleja, G.; Garcia, A.; Miravalles, R.; Galindo, T. Production of biofuels via the catalytic cracking of mixtures of crude vegetable oils and nonedible animal fats with vacuum gas oil. *Energy Fuel* **2010**, *24*, 707–717.

17. Maki-Arvela, P.; Kubickova, I.; Snare, M.; Eranen, K.; Murzin, D.Y. Catalytic deoxygenation of fatty acids and their derivatives. *Energy Fuel* **2007**, *21*, 30–41.

18. Maki-Arvela, P.; Rozmyszowicz, B.; Lestari, S.; Simakova, O.; Eranen, K.; Salmi, T.; Murzin, D.Y. Catalytic deoxygenation of tall oil fatty acid over palladium supported on mesoporous carbon. *Energy Fuel* **2011**, *25*, 2815–2825.

19. Murata, K.; Liu, Y.; Inaba, M.; Takahara, I. Production of synthetic diesel by hydrotreatment of Jatropha oils using Pt-Re/H-ZSM-5 catalyst. *Energy Fuel* **2010**, *24*, 2404–2409.

20. Wildschut, J.; Mahfud, F.H.; Venderbosch, R.H.; Heeres, H.J. Hydrotreatment of fast pyrolysis oil using heterogeneous nobel-metal catalysts. *Ind. Eng. Chem. Res.* **2009**, *48*, 10324–10334.

21. Sotelo-Boyás, R.; Liu, Y.; Minowa, T. Renewable diesel production from the hydrotreating of rapeseed oil with Pt/Zeolite and NiMo/$Al_2O_3$ catalysts. *Ind. Eng. Chem. Res.* **2011**, *50*, 2791–2799.

22. Huber, G.W.; O'Connor, P.; Corma, A. Processing biomass in conventional oil refineries: Production of high quality diesel by hydrotreating vegetable oils in heavy vacuum oil mixture. *Appl. Catal. A* **2007**, *329*, 120–129.

23. Liu, Y.; Sotelo-Boyás, R.; Murata, K.; Minowa, T.; Sakanishi, K. Hydrotreatment of Jatropha oil to produce green diesel over trifunctional Ni-Mo/$SiO_2$-$Al_2O_3$ catalyst. *Chem. Lett.* **2009**, *38*, 552–553.

24. Liu, Y.; Sotelo-Boyás, R.; Murata, K.; Minowa, T.; Sakanishi, K. Hydrotreatment of vegetable oils to produce bio-hydrogenated diesel and liquefied petroleum gas fuel over catalysts containing sulfided Ni-Mo and olid acids. *Energy Fuel* **2011**, *25*, 4675–4685.

25. Nagai, M.; Koizumi, K.; Omi, S. $NH_3$-TPD and XPS studies of Ru/$Al_2O_3$ catalyst and HDS activity. *Catal. Today* **1997**, *35*, 393–405.

26. Mott, C.J.B. Clay minerals—An introduction. *Catal. Today* **1988**, *2*, 199–208.

27. Kloprogge, J.T. Synthesis of smectites and porous pillared clay catalysts: A review. *J. Porous Mater.* **1998**, *5*, 5–41.

28. Martinez-Oritz, M.J.; Fetter, G.; Dominguez, J.M.; Melo-Banda, J.A.; Ramos-Gomez, R. Catalytic hydrotreating of heavy vacuum gas oil on Al- and Ti-pillared clays prepared by conventional and microwave irradiation methods. *Micropor. Mesopor. Mater.* **2003**, *58*, 73–80.

29. Liu, Y.; Murata, K.; Okabe, K.; Inaba, M.; Takahara, I.; Hanaoka, T.; Sakanishi, K. Selective hydrocracking of Fischer-Tropsch waxes to high-quality diesel fuel over Pt-promoted polyoxocation-pillared montmorillonites. *Top. Catal.* **2009**, *52*, 597–608.

30. Bunch, A.Y.; Wang, X.; Ozkan, U.S. Hydrodeoxygenation of benzofuran over sulfided and reduced Ni-Mo/γ-$Al_2O_3$ catalyst: Effect of $H_2S$. *J. Mol. Catal. A* **2007**, *270*, 264–272.

31. Kloprogge, J.T.; Welters, W.J.J.; Booy, E.; de Beer, V.H.J.; van Santen, R.A.; Geus, J.W.; Jansen, J.B.H. Catalytic activity of nickel sulfide catalysts supported on Al-pillared montmorillonite for thiophene hydrodesulfurization. *Appl. Catal. A* **1993**, *97*, 77–85.

32. Iranmahboob, J.; Toghiani, H.; Hill, D.O. Dispersion of alkali on the surface of Co-MoS2/clay catalyst: A comparison of K and Cs as a promoter for synthesis alcohol. *Appl. Catal. A* **2003**, *247*, 207–218.

33. Liu, Y.; Murata, K.; Hanaoka, T.; Inaba, M.; Sakanishi, K. Syntheses of new peroxo-polyoxometalates intercalated layered double hydroxides for propene epoxidation by molecular oxygen in methanol. *J. Catal.* **2007**, *248*, 277–287.

34. Liu, Y.; Murata, K.; Sakanishi, K. Hydroisomerization-cracking of gasoline distillate from Fischer–Tropsch synthesis over

bifunctional catalysts containing Pt and heteropolyacids. *Fuel* **2011**, *90*, 3056–3065.

35. Liu, Y.; Koyano, G.; Misono, M. Hydroisomerization of *n*-hexane and *n*-heptane over platinum-promoted $Cs_{2.5}H_{0.5}PW_{12}O_{40}$ (Cs2.5) studied in comparison with several other solid acids. *Top. Catal.* **2000**, *11*, 239–246.

36. Liu, Y.; Na, K.; Misono, M. Skeletal isomerization of *n*-pentane over Pt-promoted cesium hydrogen salts of 12-tungstophosphoric acid. *J. Mol. Catal. A* **1999**, *141*, 145–153.]

37. Liu, Y.; Misono, M. Hydroisomerization of n-butane over platinum-promoted cesium hydrogen salt of 12-tungstophosphoric acid. *Materials* **2009**, *2*, 2319–2336.

# Application of Heterogeneous Catalysis in Small-Scale Biomass Combustion Systems

René Bindig , Saad Butt, Ingo Hartmann,
Mirjam Matthes, and Christian Thiel

DBFZ Deutsches Biomasseforschungszentrum gemeinnützige GmbH,
Torgauer Straße 116, Leipzig 04347, Germany.

## ABSTRACT

Combustion of solid biomass fuels for heat generation is an important renewable energy resource. The major part among biomass combustion applications is being played by small-scale systems like wood log stoves and small wood pellet burners, which account for 75% of the overall biomass heat production. Despite an environmentally friendly use of renewable energies, incomplete combustion in small-scale systems can lead to the emission of environmental pollutants as well as substances which are hazardous to health. Besides particles of ash and soot, a wide variety of gaseous substances can also be emitted. Among those, polycyclic aromatic hydrocarbons (PAH) and several organic volatile and semi-volatile compounds (VOC) are present. Heterogeneous catalysis is applied for the reduction of various gaseous compounds as well as soot. Some research has been done to examine the application of catalytic converters in small-scale biomass combustion systems. In addition to catalyst selection with respect to complete oxidation of different organic compounds, parameters such as long-term stability and durability under flue gas conditions are considered for use in biomass combustion furnaces. Possible catalytic procedures have been identified for investigation by literature and market research. Experimental studies with two selected oxidation catalysts based on noble metals have been carried out on a wood log stove with a retrofit system. The measurements have been performed under defined conditions based on practical mode of operation. The measurements have shown that the catalytic flue gas treatment is a promising method to reduce carbon monoxide and volatile organic compounds. Even a reduction of particulate matter was observed, although no filtering effect could be detected. Therefore, the oxidation of soot or soot precursors can be assumed. The selected catalysts differed in their activity, depending on the compound to be oxidized. Examinations showed that the knitted wire catalyst showed better activity for the reduction of carbon monoxide, whereas the honeycomb induced a higher reduction of aromatic compounds. The properties of the two catalysts can be combined by integrating both together. The one drawback of the catalyst so far is the deactivation for the conversion of methane.

# INTRODUCTION

Due to the increase of renewable energy from small-scale biomass combustion, a reduction of flue gas emissions is necessary. Approximately 90% of renewable heat is generated by combustion of solid biomass fuels, out of which 75% is contributed by household appliances [1]. Pellet burners for the combustion of wood, as well as other alternative fuels, or manually-loaded fireplaces, release harmful and toxic emissions, which are formed during incomplete combustion. Therefore, the development of emission reduction measures such as catalytic flue gas cleaning is a necessity.

There are various organic compounds originating from biomass combustion furnaces as a result of incomplete combustion or pollutant formation processes. Among them, a large diversity of molecule structures exists, starting with several chain-like hydrocarbons, e.g., alkanes, alkenes as well as aldehydes, up to aromatic compounds like benzene and polycyclic aromatic hydrocarbons (PAH). The latter are considered as soot precursors. Furthermore, the formation of polychlorinated dibenzo-dioxins and -furans (PCDD/F) is a relevant subject in combustion research and emission reduction development, especially regarding the use of agricultural and alternative biomass fuels with a high ash-content. Emissions of several wood log stoves have been analyzed by our research group. The already published results [2] show high emissions from different kinds of wood log stoves during the start and burnout phase. In addition to gaseous organic compounds, soot and nitrogen oxides should also be reduced in flue gases. Threshold values for carbon monoxide and particulate matter in exhaust gases of small-scale firing systems are defined by the 1st Federal Emission Control Ordinance in Germany. For other compounds, no threshold values exist so far, although they are required due to their harmful properties.

The conversion of the above-mentioned organic substances to carbon dioxide and water can be accomplished by heterogeneous catalysis. Therefore, investigations have to be made about the behavior of solid and often porous catalysts in flue gases as well as reactions of gaseous compounds on the surfaces of solid catalysts.

According to the complex process of heterogeneous catalysis, one can define essential requirements regarding the use of catalysts for total oxidation in biomass combustion furnaces. First of all, a high activity along with a pronounced selectivity towards carbon dioxide is needed for the oxidation of volatile organics, carbon monoxide, and on up to PAHs. Furthermore, the activity for the oxidation of soot is also a desirable attribute and the maintenance of low nitrogen oxide emissions has to be taken into account. Moreover, a good thermal and mechanical stability of the catalyst is necessary. Lastly, the catalyst should be insensitive to poisoning, and regeneration should be possible with low energy demand as well as low costs. Taking the described characteristics into account, promising materials can be identified with respect to selected operations. One possibility is the use of supported catalysts with metal alloys (FeNiCrAl) or ceramics (cordierite, silicon carbide) as support. A porous washcoat layer, mostly consisting of -alumina, is added on the support to increase the surface area. In the porous washcoat the catalytic active species are incorporated, e.g., by impregnation. An alternative is the application of bulk or monolithic catalysts, which may consist of metal oxides or mixed metal oxides. Known substances for catalytic oxidation are noble metals: specifically platinum, palladium, ruthenium and rhodium as well as metal oxides, e.g., from copper, manganese, vanadium and tungsten, lanthanum in perovskites or spinel structures. A proper selection of catalytic substances, support and design of catalyst applications is extremely important to achieve the required operation and prevent failure due to poisoning, thermal deactivation, blockage of pores and/or active sites.

In literature, one can find some research projects of other research groups as well as developments in the commercial market, which deal with, and accordingly offer, catalytic emission reduction systems for combustion furnaces. In several papers, the results of investigations for noble metal catalysts are presented. Ahlström-Silversand *et al.* [3] and Carnö *et al.* [4] both examined the effect of noble metals compared to metal oxides. The former found that $V_2O_5/CuO$ is active for soot oxidation and palladium and platinum show good activity for high levels of conversion of volatile organic compounds (VOCs). The catalytic substances were dispersed on a -alumina coated wire mesh. Likewise, Carnö [4] concluded that platinum is more active than manganese and copper oxides on -alumina for oxidation of VOCs. In addition, the higher stability of methane compared to other VOCs, and therefore the necessity of a higher temperature for oxidation, is described on the basis of experimental data. Ferrandon and coworkers [5] investigated metallic monoliths made of iron, chromium and aluminum combined with an alumina washcoat doped with lanthanum. Manganese oxide or manganese oxide mixed with platinum were used as catalytic active substances. To carry out the activity test, a mixture of carbon monoxide, methane and naphthalene was utilized. High concentrations of manganese oxides favored the oxidation of carbon monoxide and naphthalene while lower concentrations showed higher activity for the oxidation of methane.

In Germany, the companies more Cat GmbH and Dr. Pley Environmental offer retrofit devices for emission reduction in stoves. The *"Ofenkatalysator"* (more Cat GmbH) [6] consists of noble metals (Pd, Pt) dispersed on a metallic swarf support. The needed light-off temperature is 450 K. The "ChimCat" produced by Dr. Pley Environmental has a ceramic support [7]. The catalytic active components are also noble metals, but detailed information is not given. In the USA, the integration of catalysts in the post-combustion chamber is state of the art. For instance, the company, Applied Ceramics, offers a catalyst named "Firecat," consisting of noble metals (Pt, Pd, Rh) on a ceramic honeycomb as support [8]. The light-off temperature is about 530 to 570 K. Vermont Castings have a so-called "catalytic stove" in their product line. In the post combustion chamber, a catalyst from Südchemie called "EnviCat" is integrated with noble metals as active species on a ceramic support (honeycomb). Apart from that, a similar catalyst is available from a French company named "Zéro CO Technology", which is composed of noble metals doped with cerium on a ceramic (cordierite) or metallic (FeCr alloy) support. The light-off temperature also lies between 530 and 570 K. Finally, there is also a catalytic system for nitrogen oxide reduction combined with particle filtration, which is described by Heidenreich *et al.* [9]. The equipment from Pall Filter systems GmbH includes silicon carbide candles with $V_2O_5$-$WO_3$-$TiO_2$ catalysts and a mullite membrane layer with a thickness of 150 to 200 µm. Pilot tests have been carried out at a 3.5 MW biomass combustion plant.

As shown by the examples listed above, noble metals dispersed on -alumina washcoats are frequently applied as catalysts for emission reduction. Metal oxides and mixed metal oxides are also discussed, especially for soot and tar oxidation. Nevertheless, there are still many investigations to be made for the proper use of catalysts in small-scale biomass combustion systems. Precise information about deactivation and poisoning by inorganic trace substances are required and more data about operation in real combustion furnaces is necessary.

For use in firing systems, a long-term operation is needed in terms of ecological, economical and user-friendly aspects. Appropriate solutions have to be found with respect to the design of firing systems to establish the use of catalytic treatment of flue gases from biomass combustion. Beside this, one has to pay attention to several process conditions, for example the temperature range during the burning cycle, flow rate and pressure drop as well as the durability of the materials used. The investigations carried out in our research group focus on the identification of solutions for practical realization using commercially available products. Therefore, the proportion between the costs of the firing system and the catalytic reduction system has to be considered. The aim is the development of a marketable product in the near future. By analyzing various commercially available catalysts, their performance and drawbacks are determined, to rate their applicability for flue gas treatment and to find clues for further catalyst modification or development. To consider the differences of existing furnaces, experiments are performed with different fireplaces and pellet burners. In the following sections, the investigations and results of one selected firing system are presented.

# RESULTS AND DISCUSSION

## Combustion Process and Characteristics of the Examined Wood Log Stove

Due to the different phases of a burning cycle, the conditions at the catalyst undergo cyclical fluctuations. The burning cycle includes the following phases: heating up, high temperature combustion and smoldering. The data of the reference experiment without catalyst show the characteristic emission curve for combustion processes in fireplaces. As shown in Figure 1 for carbon monoxide (CO) and total hydrocarbons (THC), the highest concentrations of the examined compounds are emitted during the heating up phase after refilling the stove. The peak concentration of CO is around 16,500 ppm and about 14,000 ppm for THC. After the high temperature combustion phase, where the emissions are comparatively low for CO (250–500 ppm) and THC (50–200 ppm), the smoldering phase follows. An increase of the CO concentration takes place up to values around 3000 ppm. During all burning phases, an excess of oxygen was observed in the flue gas. The oxygen concentration varied between 6 and 18 vol.-%, according to the burning phase, see Figure 2. Particularly, a reduction of incompletely oxidized carbon emissions was achieved during the heating up and smoldering phases by using catalysts.

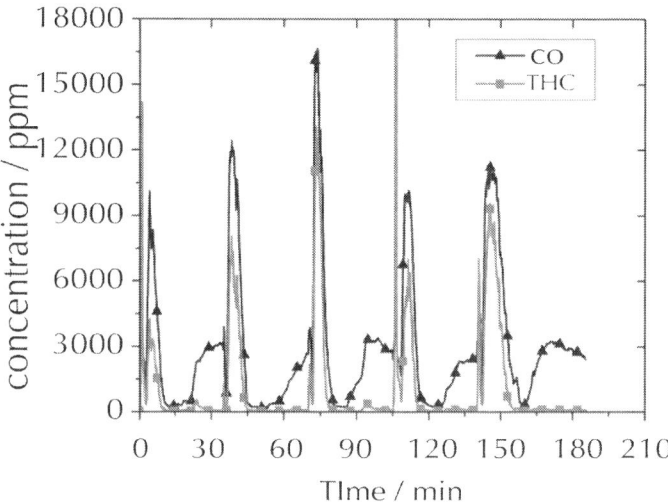

**Figure 1:** Carbon monoxide (CO) and total hydrocarbon (THC) concentration in the exhaust gas during several burning cycles in the wood log stove used during experimentation.

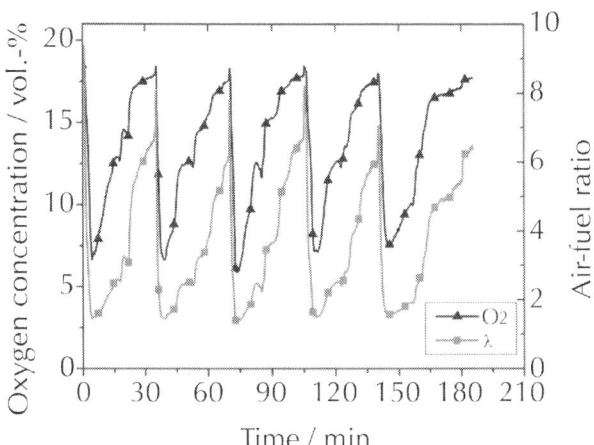

**Figure 2:** Oxygen concentration and ratio of actual air-fuel ratio to stoichiometric ratio ( ) in the exhaust gas during several burning cycles in the wood log stove.

For the experiments described in this paper, neither temperature nor volumetric flow showed any significant differences. Therefore, average values determined from all experiments are presented. The conditions at the catalyst undergo alterations due to the cyclic combustion process. The temperatures at the catalyst vary from 570 to 830 K. The highest temperatures occur during the high temperature combustion phase. Figure 3 shows the temperature curve for five burning cycles.

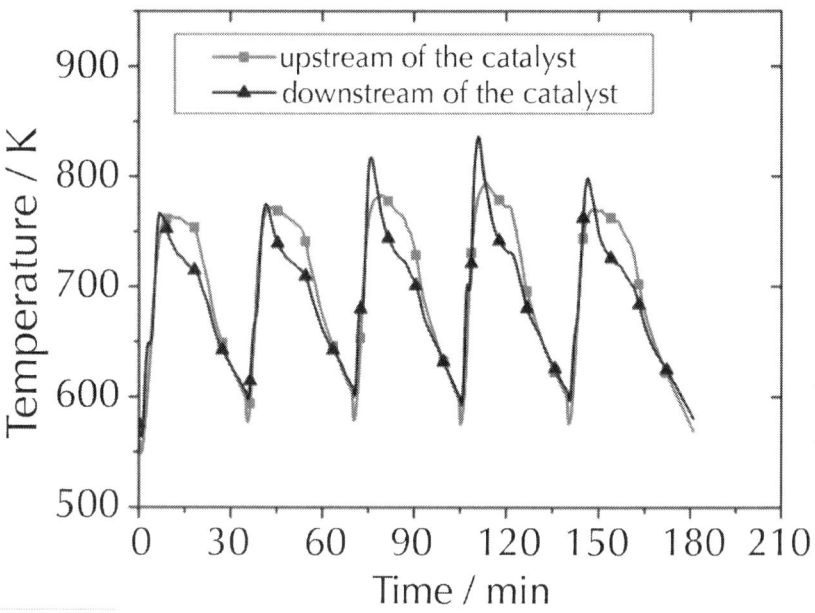

**Figure 3:** Temperature upstream and downstream of the catalyst during several burning cycles in the wood log stove.

The volumetric flow through the firing system is dependent on the chimney fan, whose power is regulated to achieve a constant negative pressure inside of the combustion chamber, upstream of the catalyst. Therefore, the highest flow occurs during the high temperature combustion phase, when a considerable amount of gas is released. Figure 4 shows the curve of the volumetric flow during five burning cycles. The pressure drop across the catalyst also varies, as it is dependent on the volumetric flow. The residence time of the flue gas in the catalyst fluctuates as presented in Figure 5. For one honeycomb catalyst unit the residence time of the flue gas is between

0.015 and 0.030 s, for two catalyst units the range is between 0.035 and 0.055 s. The volume of the knitted wire catalyst unit is about 85% of the honeycomb. Accordingly, the residence time is 85% of the value presented for the honeycomb.

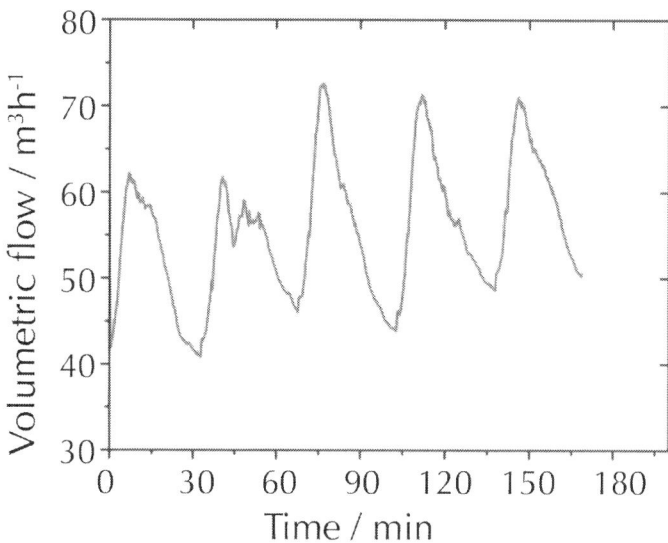

**Figure 4:** Volumetric flow through the firing system and accordingly through the catalyst in the wood log stove.

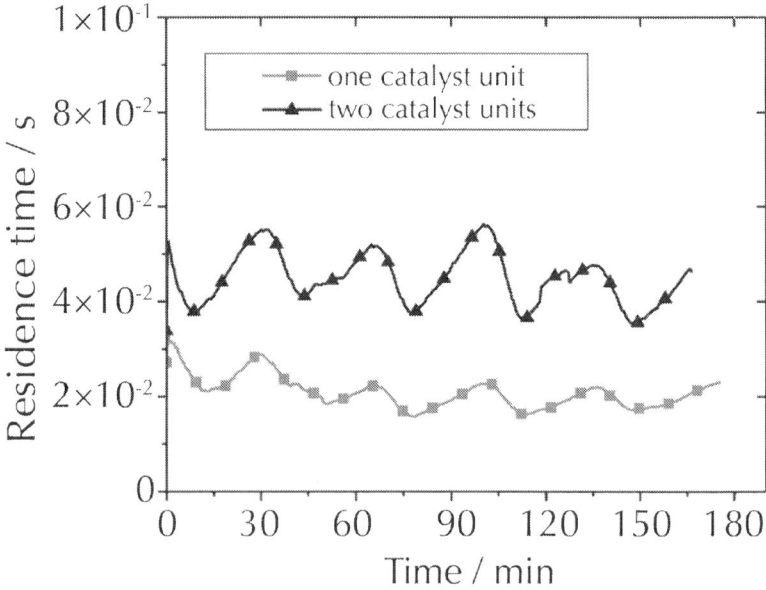

**Figure 5:** Residence time of the flue gas in the catalyst during operation with one and two catalyst units. (The values are not dependent on the different materials used.)

A temperature-dependent conversion or reduction rate analysis is not reasonable so far, because the stove can only operate under transient conditions with fast changes in pollutant concentrations and temperature. As the catalysts were not operated under isothermal conditions, it is not possible to get reliable data with the used experimental setup. Therefore, mean values of five complete burning cycles were calculated for each experiment to analyze the pollutant reduction behavior of the different catalysts under examination. In future experiments, the recording of three-dimensional temperature profiles of catalysts should be carried out. Also, simultaneous gas phase measurements with FTIR are necessary, upstream and downstream of the catalyst. Furthermore, the experimental data should be compared with reactor model simulation.

# Catalytic Flue Gas Cleaning in a Retrofit System in a Wood Log Stove

## *Activity and Pressure Drop According to Material and Catalyst Volume*

Both tested catalysts, the ceramic honeycomb, as well as the knitted wire catalyst, showed considerable activity for oxidation of organic flue gas components. Both were inserted fresh after manufacturing and delivery without previous usage or conditioning. Indeed, their reactivity differs for the component to be oxidized. The results of the measurements are presented in Figure 6. The knitted wire catalyst unit showed a higher reduction of carbon monoxide and higher formations of nitrogen oxides than the honeycomb catalyst, whereas the reduction of aromatic compounds of 86.3% was higher for the honeycomb than for the knitted wire (70.9%). The reduction of carbon monoxide was 99.5% using the knitted wire catalyst and 92.6% using the honeycomb catalyst. The nitrogen oxide concentration was increased to 139.1% for the knitted wire in relation to the reference experiment and to 110.0% for the honeycomb. Also, a reduction of 80% was observed for all measured hydrocarbons. For methane, the reduction was 80.4% (honeycomb) and 84.5% (knitted wire); for acetylene it was 94.5% (honeycomb) and 92.4% (knitted wire).The reduction of particulate matter (PM) emission is again higher using the knitted wire catalyst. For the knitted wire, the concentration was reduced by 49.9%, and for the honeycomb by 34.3%. Comparing the emission reduction of the catalysts, one has to take into account that the residence time is not the same for both, due to their different design. Therefore, the comparison given refers to the use of commercially-available catalytic devices under flue gas conditions and not to the catalytic activity in relation to the same catalytic surface area.

The pressure drop across the honeycomb catalyst is higher than that of the knitted wire catalyst. As both show no increasing trend over several burning cycles (see Figure 7), their use for continuous operation is promising.

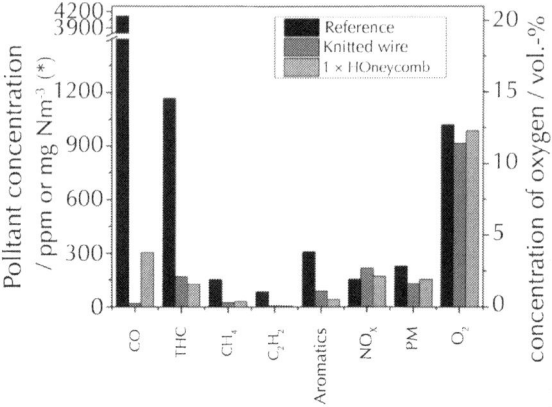

**Figure 6:** Comparison of the concentrations of selected pollutants and of oxygen in the exhaust stream without catalyst and with the two tested catalysts (black bars: without catalyst; red bars: knitted wire catalyst, green bars: honeycomb catalyst; * The measurement unit of the particulate matter concentration is mg Nm$^{-3}$ and for all other pollutants ppm).

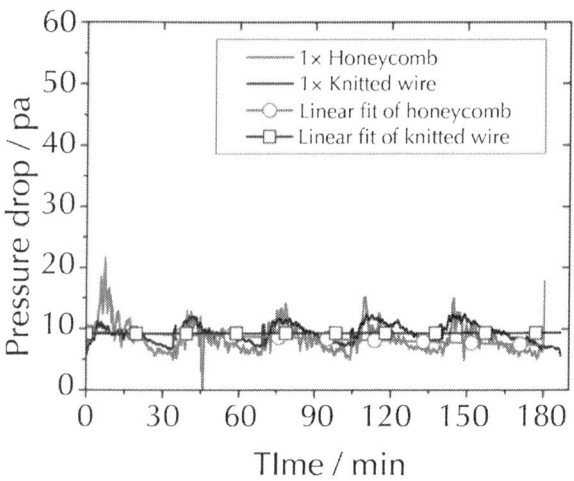

**Figure 7:** Signal curve and trend (linear fit) for the pressure drop across the catalyst during operation time.

By increasing volume of the honeycomb catalyst, the reduction of gaseous carbonaceous compounds is increased for CO to 95.3% and for THC to 90.8%. But at the same time, the concentration of nitrogen oxides is slightly higher with a concentration of 121.7% in relation to the reference value. For the other investigated carbon compounds (THC, $CH_4$, $C_2H_2$ and aromatics), only a small change of 1–4% was observed. The reason for the minor change after inserting a second catalyst compared to the difference between the reference and one catalyst unit might be given in the concentration dependency of the reaction rate. The second catalyst unit had to be placed downstream of the first unit, because of the construction of the retrofit system. Therefore, much lower pollutant concentrations were present in the second unit, obviously leading to a lower reaction rate. The particulate matter emission remains the same. As a result, no significant filter effect is observed for the catalyst. Figure 8 shows the measured emissions, comparing one and two catalyst units. Even though the range of the pressure drop is naturally increased by extension of the volume, see Figure 9, again no upward trend is observed. A secure operation seems possible.

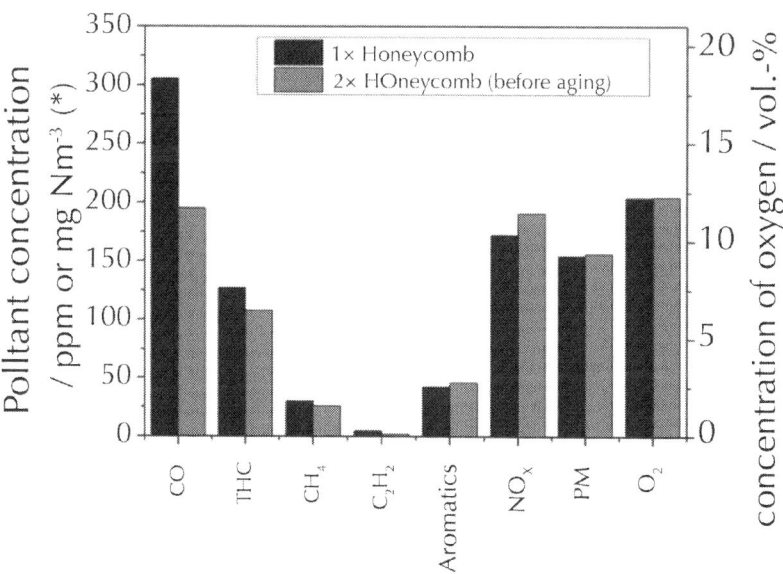

**Figure 8:** Comparison of the concentrations of selected pollutants and of oxygen in the exhaust stream for one and two catalyst units (black bars: one catalyst unit; red bars: two catalyst units. * The measurement unit of the particulate matter concentration is mg $Nm^{-3}$ and for all other pollutants ppm).

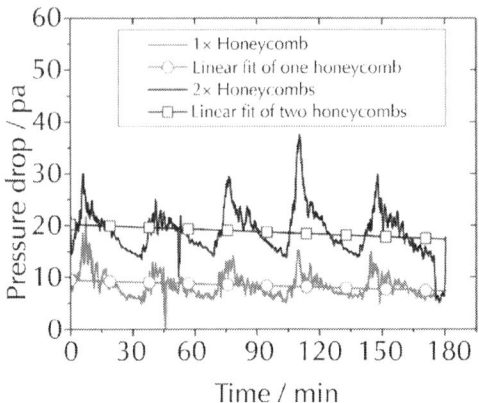

**Figure 9:** Signal curve and trend (linear fit) for the pressure drop across the one and two honeycomb catalyst units during operation time.

Ozil *et al.* [10] already studied the efficiency of catalytic processes for the reduction of emissions from small-scale combustion systems. They also used two different kinds of catalysts. One was based on cordierite support and another one on a metallic support (corrugated structure). In addition, the influence of catalyst heating was studied. The cell densities of the catalysts were quite low (25 cpsi for the cordierite support and 16 cpsi for the metallic one) to meet the requirement of a low pressure drop. A chimney fan was used in the current study, so catalysts with higher pressure drops could be tested. That is, catalysts with higher cell densities (100 cpsi for the cordierite honeycomb) or made of knitted wires. Furthermore, results concerning the long-term stability of the catalysts or a combination of both were not mentioned in [10]. The emission reductions reported by Ozil *et al.* [10] were lower (e.g., CO reduction of 80 to 90%) compared to the findings in this work (CO reduction > 90%). A reason for the differences could be the higher cell densities of the catalysts, *i.e.*, higher catalytic surface area, used in this work.

## Deactivation During Operation

The comparative measurements of the honeycomb catalyst in the retrofit system showed a slight change in the flue gas concentration of selected pollutants after two performed experiments, according to the routine described in Section 3 (Experimental Section). While the reduction of CO and aromatic compounds is even a little higher after several burning cycles than in the fresh state, the reduction of total hydrocarbons (THC), methane as well as acetylene, and also the formation of nitrogen oxides is lower. The nitrogen oxide concentration is actually 100.8% and therefore as low as during the reference examinations without catalyst. The results are represented in Figure 10. The major change of carbon emissions was observed for methane, where the reduction was decreased to 67.7%. Again, the pressure drop showed no upward trend during the experiment, even though the range was a little higher than before the two aging experiments. The particulate matter emission is the same before and after the two aging experiments.

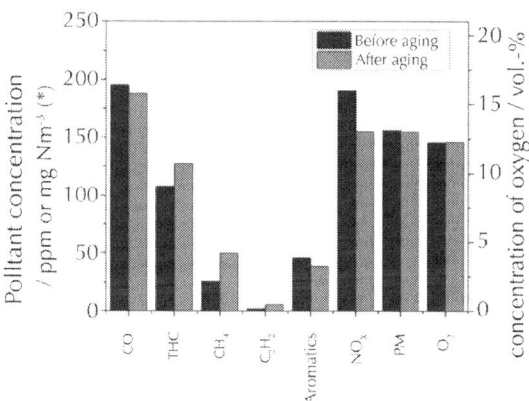

**Figure 10:** Comparison of the concentrations of selected pollutants and oxygen in the exhaust stream after the catalyst in the fresh state (black bars) and after 16 h of use (red bars); * The measurement unit of the particulate matter concentration is mg Nm$^{-3}$ and for all other pollutants ppm.

Experiments for the deactivation of the knitted wire catalyst are not carried out yet. The performance is planned when appropriate samples for analysis in the catalyst testing apparatus are received.

## Combination of Honeycomb and Knitted Wire Catalyst

Because the examined catalysts showed different behavior regarding the measured pollutants, the use of both was assumed to be a good possibility to combine their advantages. Both inserted catalytic devices had already been used in the foregoing experiments as described above. The honeycomb catalyst was placed upstream of the knitted wire catalyst during the experiment. The setup was chosen according to the less flexible form of the honeycomb, which offered a supporting surface for the knitted wire unit. In correlation with the foregoing assumption, the emission values can be attributed to some extent to the catalyst that showed a higher activity when used as a single unit. More precisely, as depicted in Figure 11, the emissions of CO are close to the values of the experiment with the knitted wire catalyst. The reduction is 98.1% in relation to the reference. The concentrations of THC as well as aromatic compounds are analogous to the experiment with the honeycomb catalyst. A reduction of 89.0% for THC and 87.3% for aromatic compounds was observed. The reduction of 93.4% for acetylene is analog to the values measured at the experiments with the single units. For interpretation of this comparison, one also has to take into account that the catalyst deactivates while using it in the retrofit system. On this basis, the slightly higher methane concentration of the combined catalysts in comparison to the values of the single ones can be explained by the deactivation of the catalysts with respect to methane oxidation. The reduction of 66.5% for methane is about the same as for the aged honeycomb catalysts. The data for the activity measurements of methane oxidation will be given in the following chapter.

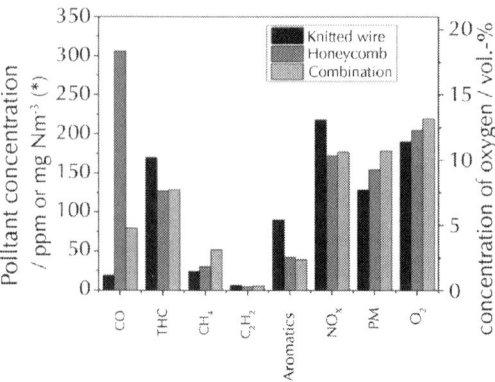

**Figure 11:** Comparison of the concentrations of selected pollutants and oxygen in the exhaust stream after the catalyst (black bars: knitted wire catalyst;

red bars: honeycomb catalyst, green bars: combination of honeycomb and knitted wire catalyst; * The measurement unit of the particulate matter concentration is mg $Nm^{-3}$ and for all other pollutants ppm).

# Catalytic Activity of Monolithic Catalysts for Methane Oxidation

By interpretation of the data from the CTA-experiments (catalyst testing apparatus), the deactivation of the honeycomb catalyst was investigated during the use in the retrofit system. Furthermore, a trend for the long-term stability can be observed. While preparing the samples of the knitted wire catalyst and their installation into the CTA, the washcoat of the catalyst was easily removed from the wire by mechanical stress. Therefore, the analysis of the knitted wire catalysts could not be carried out in a reproducible way. Hence, reliable data only exist for the honeycomb catalyst.

The examined honeycomb catalyst was in use for a total time of six experiments (procedure described in Section 3) within the self-constructed retrofit device called RFS-TKI. It was exposed to temperatures between 550 and 800 K during its use. The temperatures were subjected to strong fluctuations, see Figure 3. Fluctuations of the exhaust gas temperature and highly variable composition of flue gases are typical for manually loaded small-scale firing systems, such as wood log stoves.

The results from the investigation (temperature-conversion curves) of the honeycomb catalyst before and after usage in the retrofit system RFS-TKI are shown in Figure 12. The loss of activity for the oxidation of methane after usage in the retrofit system is obvious. In the fresh state, a methane conversion of 85% at 800 K was recorded. After usage in the firing system, the conversion was only 62% at the same temperature. Given that the concentration of other carbonaceous compounds than methane, like CO and aromatics, was not higher after several hours of catalyst usage, a selective or specific deactivation can be assumed, which especially affects the catalytic methane oxidation process.

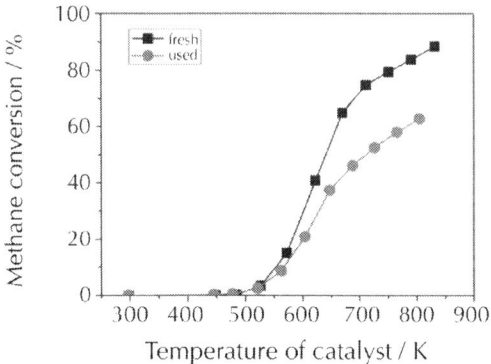

**Figure 12:** Illustration of the loss of activity towards the oxidation of methane after usage in the retrofit system RFS-TKI by means of the temperature-conversion behavior.

Methane was used as a model flue gas compound, as its stability is known to be high. The critical step for the oxidation of methane is the cracking of the CH-bond. The bonding strength requires high activation energy for the reaction. If total oxidation of methane (see overall reaction below) is achieved by the catalyst, it is likely that a wide range of other carbon compounds can also be oxidized.

$$CH_4 + 2O_2 \xrightarrow{k_{eff}} CO_2 + 2H_2O$$

The Arrhenius (Figure 13) plots were calculated from the temperature-conversion curves of both catalyst samples for interpretation of the experimental data. The quantification of the activity of the catalysts was analyzed by determining the pre-exponential factors and the activation energies. To calculate the rate constants ($k_{eff}$) at the different temperatures, the following assumptions were made. The reaction is pseudo first order, because of the high oxygen excess. It is not reversible and takes place without volume change. Temperature gradients within the catalyst are negligible.

Furthermore, the following rate equation was used:

$$\frac{dc}{dt} = -k_{eff} \cdot c$$

By using a rate constant related to the catalyst mass ($k_{eff}^m$),

$$k_{eff}^m = \frac{k_{eff}}{m_{cat}}$$

the rate equation changes to:

$$\frac{dc}{dt} = -k_{eff}^m . m_{cat} . c$$

By separation of the variables and using the integration limits ($c_0$ = concentration at reactor inlet, $c_A$ = concentration at reactor outlet, $\tau$ = residence time) $c(t = 0) = c_0$ and $c(t = \tau) = c_A$ with ($\dot{V}_R$ = volumetric flow at reaction conditions) $\tau = \frac{V_R}{\dot{V}_R}$ the following expression was obtained:

$$\ln\left(\frac{c_0}{c_A}\right) = k_{eff}^m . m_{cat} . \frac{V_R}{\dot{V}_R}$$

After inserting

$$\frac{c_0}{c_A} = \frac{1}{1-U} \quad \text{and} \quad \dot{V}_R = \frac{p_1 . \dot{V}_1 . T_R}{T_1 . p_R}$$

and rearranging, the following equation was obtained:

$$k_{eff}^m = \ln\left(\frac{1}{1-U}\right) . \frac{p_1 . \dot{V}_1 . T_R}{T_1 . p_R . m_{cat} . V_R}$$

All concentrations used to calculate the conversion ($U$) were average values over 5 min, which were determined after a stable concentration level was achieved.

In the lower temperature range, the plot yields a straight line for both catalysts, as can be seen on the right hand side of Figure 13. This implies that the assumption of a pseudo first order reaction was reasonable. At higher temperatures (above 700 K), there were also straight lines, but with a smaller absolute value of the slope. The rate constants, which result from the right part of the plot, were nearly twice as high as those from the left part. This observation can be explained by the occurrence of mass transfer limitation. Therefore, the left part of the plot was not considered further.

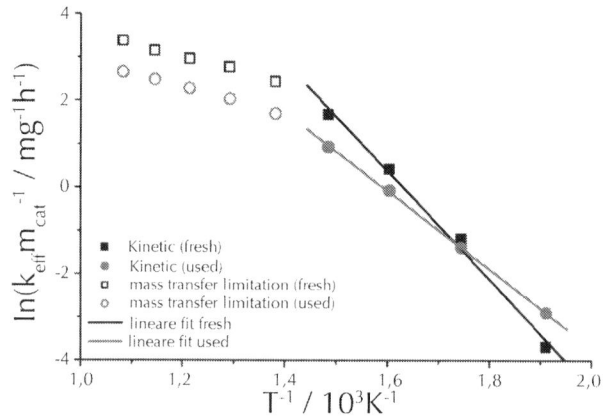

**Figure 13:** Arrhenius plots of honeycomb catalysts (fresh and used).

The experimental parameters, pre-exponential factors and the apparent activation energies (calculated from the right part of the Arrhenius plot) for both catalysts are listed in Table 1. The pre-exponential factor was reduced after usage in the RFS-TKI. This finding can be explained by loss of catalytic surface area due to the formation of larger noble metal crystallites out of smaller ones. The fluctuations in the exhaust gas temperature represent a particularly high stress on catalysts and can lead to a significant reduction of the active surface.

The apparent activation energy of the used catalyst was also reduced. One explanation for this is the formation of PdO from Pd during the stove experiments. Pd/PdO is believed to be an active species for the total oxidation of methane [11].

**Table 1:** Experimental parameters of both catalysts

| Parameter | Symbol | Unit | Fresh honeycomb | Used honeycomb |
|---|---|---|---|---|
| Catalyst mass | $m_{cat}$ | g | 8.2 | 7.8 |
| Catalyst diameter | $r$ | m | 0.0027 | 0.0027 |
| Catalyst length | $l$ | m | 0.0024 | 0.0023 |
| Catalyst volume | $V_R$ | m³ | $1.37 \times 10^{-5}$ | $1.32 \times 10^{-5}$ |
| Standard volumetric flow | ■ | m³ h⁻¹ | 0.501 | 0.501 |
| Activation energy | $E_A$ | kJ mol⁻¹ | 104 | 75 |
| Pre-exponential factor | $k_{effm,0}$ | h⁻¹ mg⁻¹ | $7.756 \times 10^8$ | $1.778 \times 10^6$ |

In addition to the permanent blocking of active sites by gaseous components (catalyst poisoning), either liquid or solid particulate matter can also lead to a reversible blocking of active sites (clogging). A loss of active surface or the blocking of active sites leads to a reduction in the activity of the catalyst. To understand the exact mechanism of the deactivation, more detailed studies are planned.

A further target of catalyst analysis is the determination of the activity in relation to the inner surface of the active species or catalyst mass to obtain conclusions regarding the most efficient catalyst composition. Then, an assessment can be given with respect to the efficient use of resources, as well as economic aspects. Therefore, it is intended to further cooperate and interact with catalyst manufacturers and researchers dealing with catalysis.

# EXPERIMENTAL SECTION

For the testing of catalysts regarding their behavior and effect in biomass combustion flue gases, retrofit systems combined with several wood log stoves as well as small pellet burners are used at the *Deutsches Biomasseforschungszentrum* (DBFZ). In addition, a catalyst testing apparatus is available to investigate probes under defined conditions with selected gases, like methane and propane. The intent behind the work is to develop practical solutions. Therefore, the focus is on the analysis of real conditions and the measurement of a number of relevant compounds. A detailed investigation of the overall reactions that occur cannot be undertaken, due to the complex system, which includes several decomposition, oxidation and pollutant formation mechanisms.

## Retrofit System and Wood Log Stove

The results described were obtained during the operation with a wood log stove in combination with a retrofit system. A double-walled wood log stove "Zeus/Odin" from the brand CAMINOS with a heat output of 7 kW was used, which was purchased from a local construction store, *i.e.*, it is a stove in the lower price category. The self-constructed retrofit system is used for the catalytic treatment of flue gases from wood log stoves at the DBFZ. The bypass can ensure safe operation, even in the case of catalyst blocking. The system includes a heat transfer zone to increase the heat output and is called RFS-TKI. In Figure 14, the setup of the system is illustrated. Furthermore, another retrofit system named "KATI" exists at the DBFZ for which results have already been published in [2].

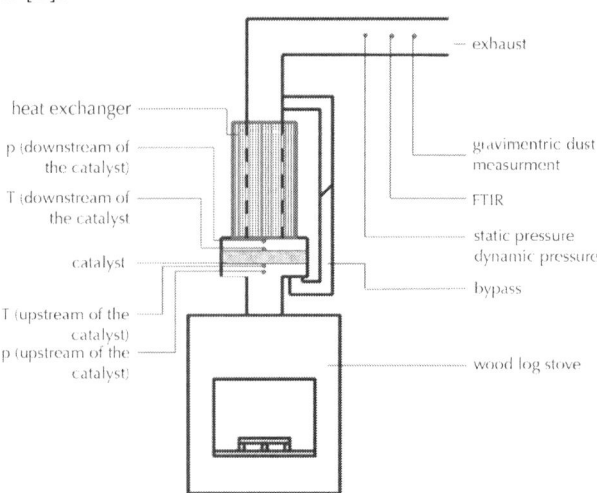

**Figure 14:** Schematic illustration of the combination of wood log stove and retrofit system RFS-TKI.

Standardized wood logs [12] were used as fuel during the experiments. The measurement of temperature and pressure inside the retrofit system and flue gas pipes were carried out with type K thermocouples and differential pressure sensors (type SDP 1000-L025, Manufacturer: SENSIRION) combined with Prandtl probes. The concentration of the released gaseous emissions: CO, $CH_4$, $NO_x$ and $SO_2$, as well as VOCs like formaldehyde or BTX, were continuously measured using Fourier transform infrared spectroscopy (FTIR, Manufacturer: Calcmet). Also, a paramagnetic Oxygen-Analyzer (Manufacturer: M&C, type: PMA 100) and a flame ionization detector (FID, Manufacturer: Mess- & Analysentechnik GmbH, type: Thermo-FID ES) were used to measure oxygen and total volatile organic carbon compounds. The total particulate matter concentration in the flue gas was determined according to the VDI 2066/1 with gravimetric measurement.

For interpretation of the experiments, the following compounds and sum parameter given in Table 2 were analyzed. The chosen substances are relevant, due to their role in the combustion process or pollutant formation mechanism, as well as their environmental effects and other effects harmful to health. The concentration of carbon monoxide and oxygen are indicators of combustion quality. The sum parameter "aromatics" includes several compounds with one or two aromatic rings, such as benzene, acenaphthene and naphthalene. Aromatic and aliphatic compounds like acetylene are believed to be precursors of soot. The analysis of the particulate matter concentration is important in respect of two potential emerging processes. One is a possible reduction of PM that can be achieved by oxidation of soot particles as well as precursors of soot. The second is a possible filter effect of catalyst units. This second one is an unwanted effect, because it can lead to blockage of pores and active sites of the catalyst.

**Table 2:** Analyzed compounds and sum parameters

| Carbon monoxide (CO) |
| --- |
| Total hydrocarbons (THC) |
| Methane ($CH_4$) |
| Acetylene ($C_2H_2$) |
| Aromatics |

| Nitrogen oxides ($NO_x$) |
| Oxygen ($O_2$) |
| Particulate matter (PM) |

All experiments were performed according to a defined routine to generate comparable data. The experimental routine was defined to measure application-oriented conditions for stove operation. At first, two burning cycles were performed to ignite the fire and heat up the stove. Then five burning cycles followed, which were used to measure the described process parameters and flue gas components. One burning cycle lasts for 35 min and may be sub-divided into the following three phases: heating up (10 min), high temperature combustion (10 min) and smoldering (15 min). Total flame extinction and cooling of the stove in the smoldering phase is avoided by the given routine, because it leads to an ineffective state, where high pollutant concentrations are emitted. Average values of the burning cycles were determined for interpretation and comparison of the experiments. The given data in the chapter "Results and discussion" include average values of all five burning cycles. The concentrations are based on the natural oxygen content and dry state of the exhaust gas.

Two different catalyst types were tested in the experiments with the wood log stove, which primarily differ in their supporting material. More precisely, honeycomb and knitted wire supports have been tested. The applied noble metal catalysts, out of Pt and Pd on -$Al_2O_3$, were provided by Heraeus Precious Metals GmbH & Co. KG. Further detailed information about the catalysts is not disclosed, in accordance with the cooperation agreement. The honeycomb support consists of cordierite. The knitted metal wire is made out of special high alloy steel.

For installation in the RFS-TKI, the catalyst units were wrapped with a high temperature resistant felt consisting of $SiO_2$, $Al_2O_3$ and $ZrO_2$ to seal the space between catalyst and flue gas pipe. The size of the two different catalyst units was determined, according to the generated pressure drop. Primarily, the secure use of catalysts is limited by their pressure drop, since an excess pressure in the firing systems has to be prevented.

The influences of volume as well as activity loss during several burning cycles of the honeycomb catalysts were examined. For comparison, the operation of the knitted wire catalyst plus the combination of honeycomb and knitted wire catalyst were examined. The emissions of the stove without using a catalyst were measured as reference values. In Table 3 the different versions examined are listed.

**Table 3:** Overview of the different versions for the examination

| Version | Catalyst arrangement for experiment |
|---------|-------------------------------------|
| 1 | Reference without catalyst |
| 2 | One fresh knitted wire unit |
| 3 | One fresh honeycomb unit |
| 4 | Two honeycomb units (fresh) |
| 5 | Two used honeycomb units (both used before the experiment for 14 burning cycles, equivalent 490 min) |
| 6 | Combination of one knitted wire and one honey-comb unit (both used before for 7 burning cycles, equivalent 245 min) |

# Catalyst Testing Apparatus

The catalyst testing apparatus (CTA) comprises equipment used for the comparison and quantification of the activity of catalysts. Basically, the CTA consists of a fused silica reactor [13], a tube furnace, a connection for test gas cylinders, a connection for compressed air, mass flow controllers, a gas inlet pipe system and a 1311 Fast Response Triple-gas Monitor (FRTM). The FRTM allows the simultaneous measurement of carbon dioxide, oxygen and a specified hydrocarbon concentration in a gas flow. The measurement principles are based on photo-acoustic spectroscopy (for carbon dioxide and hydrocarbons) and magneto-acoustic spectroscopy (for oxygen).

Figure 15 shows a schematic drawing of the CTA. The gas inlet pipe system is built up of PFA-tubes coupled by compression fittings out of stainless steel. The mass flow controllers are purchased from Bronkhorst and the FRTM from Innova AirTech Instruments. Cylindrical units with a diameter of 27 mm have been removed from the catalysts for installation of the samples under examination in the testing apparatus. To prevent any bypass flow, a high temperature resistant felt, consisting of $SiO_2$, $Al_2O_3$ and $ZrO_2$, was used to seal the space between the samples and the reactor wall. The volumetric gas flows of air and methane were kept constant for all experiments. The methane concentration was adjusted to 1500 ppm. The procedure on the CTA was implemented in line with the experiments on the retrofit systems. A comparison with reference to the catalyst volume is possible, considering the catalyst as a unit consisting of support, washcoat and catalytic active substance.

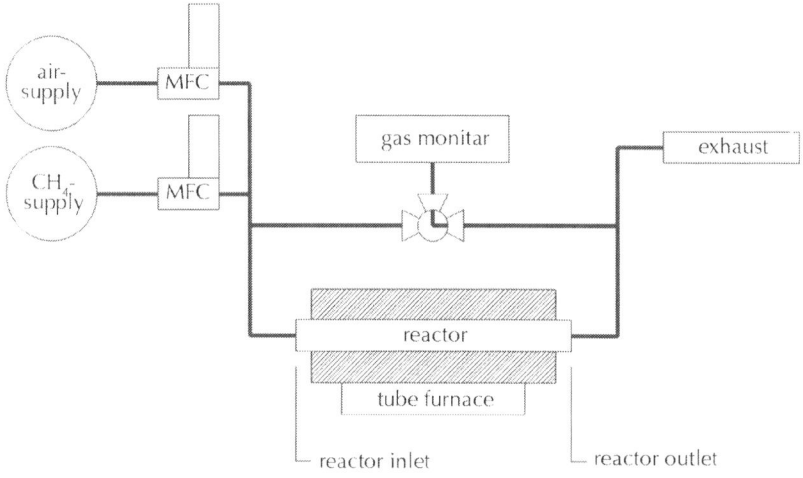

**Figure 15:** Schematic illustration of the catalyst testing apparatus.

# CONCLUSIONS AND OUTLOOK

The results obtained with the selected catalysts in a wood log stove demonstrate their potential for emission reduction. In particular, a significant reduction of carbon monoxide and organic carbon is possible. Even the emission of carbon compounds with a more stable structure like aromatics and methane can be reduced to some extent. The highest reduction of 99.5% for CO was observed using the fresh knitted wire catalyst. Also, the highest nitrogen oxide concentration was measured for this setup. Contrary to these results, higher conversions of aromatic and total hydrocarbon compounds were obtained by using the honeycomb catalyst. The emission reduction was 89.1% for total hydrocarbons and 86.3% for aromatics in the experiment with a single fresh unit. Only the reduction of methane and the formation of nitrogen oxides decreased during the performed experiments, therefore indicating a deactivation of the catalytic material. The decline of the increased emission of nitrogen oxides with catalysts observed after several hours of use can be stated certainly as a positive effect of deactivation. The activity for catalytic oxidation of carbon monoxide as well as aromatics appears not to be affected from the deactivation process observed for methane.

The analysis of the honeycomb catalyst samples with the catalyst testing apparatus confirm the results with the wood log stove and retrofit system. The deactivation of the catalytic activity for the oxidation of methane is shown after short use in biomass flue gas. In the process, the effect of decreasing activation energy after usage does not weigh the loss of catalytic surface area. According to the results, a differentiated analysis has to be done on deactivation processes of catalysts. Therefore, the use of other model gases, such as propane and carbon monoxide is planned with the catalyst testing apparatus.

However, in terms of the implementation of a comprehensive solution for catalytic emission reduction in biomass flue gases, no satisfying long-term stability can be assured so far for the catalysts under analysis. It has to be mentioned, though, that the samples are not specifically fabricated for use in biomass flue gases. In this context, the influence of a pretreatment for the catalysts is relevant. It has to be analyzed if a stable activity level is reached after a certain extent of deactivation, or if a total activity loss is reached after several hours of operation. Finally, the investigations have to be extended to intervals related to the heating periods of firing systems.

Currently, and in the future, analysis of other supporting materials and also other catalytic substances are being (will be) carried out to identify the most suitable catalyst arrangement. Metal foams can be an alternative to the analyzed supporting materials used so far, and metal oxides have the potential to replace noble metal-based catalysts. Lower costs of a catalytic device may be achieved with the use of metal oxides, considering the market price of noble metals. Investigations have to be carried out regarding the catalytic active surface and the amount of the used catalytic material per unit volume of the catalyst to provide a more comprehensive comparison of the different materials.

# ACKNOWLEDGMENTS

This article was written in context of the competence group "Catalytic Emissions Reduction" at the DBFZ (*Deutsches Biomasseforschungszentrum gemeinnützige GmbH*), supported by the *Bundesministerium für Ernährung, Landwirtschaft und Verbraucherschutz* of the Federal Republic of Germany.

# REFERENCES

1.  Lenz, V.; Kaltschmitt, M. Erneuerbare Energien. *BWK Das Energie-Fachmagazin* 2011, *63*, 42–55.

2.  Bindig, R.; Hartmann, I.; Koch, C.; Matthes, M.; Schenker, M.; Thiel, C.; Kraus, M.; Roland, U.; Einicke, W.-D. Abgasreinigung an Biomassekleinfeuerungsanlagen und experimentelle Untersuchungen zur Kombination von katalytischen und elektrostatischen Abgasreinigungsverfahren. *Chem. Ing. Tech.* 2011, *83*, 2105–2120.

3.  Ahlström-Silversand, A.F.; Odenbrand, C.U.I. Thermally sprayed wire-mesh catalysts for the purification of flue gases from small-scale combustion of bio-fuel Catalyst preparation and activity studies. *Appl. Catal. A* 1997, *153*, 177–201.

4.  Carnö, J.; Berg, M.; Järas, S. Catalytic abatement of emissions from small-scale combustion of wood: A comparison of the catalytic effect in model and real flue gases. *Fuel* 1996, *75*, 959–965.

5.  Ferrandon, M.; Berg, M.; Björnbom, E. Thermal stability of metal-supported catalysts for reduction of cold-start emissions in a wood-fired domestic boiler. *Catal. Today* 1999, *53*, 647–659.

6.  moreCat GmbH. Metal Catalysts. Available online: http://www.morecat.de/start%20EN/index.html (accessed on 7 December 2011).

7.  Dr. Pley Environmental GmbH. Saubere Energie aus Biomasse. Available online: http://www.chimcat.com/chimcat (accessed on 7 December 2011).

8.  Gs-components handelsgesmbH. Firecat. Available online: http://www.firecat.cc/html/english.html (accessed on 7 December 2011).

9.    Heidenreich, S.; Nacken, M.; Hackel, M.; Schaub, G. Catalytic filter elements for combined particle separation and nitrogen oxides removal from gas streams. *Powder Technol.* 2008, *180*, 86–90.

10.   Ozil, F.; Tschamber, V.; Haas, F.; Trouvé, G. Efficiency of catalytic processes for the reduction of CO and VOC emissions from wood combustion in domestic fireplaces. *Fuel Process. Technol.* 2009, *90*, 1053–1061.

11.   Farrauto, R.J.; Hobson, M.C.; Kennelly, T.; Waterman, E.M. Catalytic chemistry of supported palladium for combustion of methane. *Appl. Catal. A* 1992, *81*, 227–237.

12.   Hartmann, I.; Lenz, V.; Schenker, M.; Thiel, C.; Roland, U.; Einicke, W.-D.; Bindig, R. Katalysatoruntersuchungen an einer Technikumsanlage für Biomasse-Kleinfeuerungen. *Chem. Ing. Tech.* 2011, *83*, 1–7.

13.   Bindig, R.; Hartmann, I.; Thiel, C. Apparatus for identification of applicable catalysts for exhaust gas purification at small-scale biomass combustion systems. In Proceedings of the IBC LEIPZIG—International Biomass Conference, Leipzig, Germany, May 2011.

# Chapter 6

# Biomass Converting Enzymes as Industrial Biocatalysts for Fuels and Chemicals: Recent Developments

Matt D. Sweeney and Feng Xu

Novozymes Inc., 1445 Drew Avenue, Davis, CA 95618, USA

## ABSTRACT

The economic utilization of abundant lignocellulosic biomass as a feedstock for the production of fuel and chemicals would represent a profound shift in industrial carbon utilization, allowing sustainable resources to substitute for, and compete with, petroleum based products. In order to exploit biomass as a source material for production of renewable compounds, it must first be broken down into constituent compounds, such as sugars, that can be more easily converted in chemical and biological processes. Lignocellulose is, unfortunately, a heterogeneous and recalcitrant material which is highly resistant to depolymerization. Many microorganisms have evolved repertoires of enzyme activities which act in tandem to decompose the various components of lignocellulosic biomass. In this review, we discuss recent advances in the understanding of these enzymes, with particular regard to those activities deemed likely to be applicable in commercialized biomass utilization processes.

# INTRODUCTION

Converting renewable, widely available yet vastly underused lignocellulosic biomass to valuable chemicals, including fuel and polymer precursors, is of strategic importance for the sustainability and advancement of energy and chemicals industries (for recent reviews, see [1,2,3,4,5,6,7,8,9,10,11]). Compared to the commercialized starch and sugarcane-based "first generation" biofuel (starch or sucrose conversion to ethanol), emergent biomass-based "second generation" biofuel ((hemi) cellulose conversion to ethanol) has the potential of not only significantly displacing fossil fuels, but also adding value to agricultural byproducts, forestry residues, or municipal wastes. Biomass conversion is also being expanded beyond fuel production; the concept of bio refinery is being actively pursued so that a wide range of useful materials (for chemicals, energy, food, healthcare, and other industries) can be derived from biomass.

Naturally occurring lignocellulosic biomass, especially in plant cell walls, serves structural and protective roles for plants, and consequently is recalcitrant and resistant to degradation. It is a major challenge to convert or degrade at industrial scale highly complex and heterogeneous lignocellulosic biomass into simple carbohydrates, phenolics, aromatics, and other more transformable substances. Among the numerous physical, chemical, and biological methods under development, the ones relying on enzymes are particularly attractive. Natural lignocellulose degradation and utilization (as part of natural energy transfer and carbon cycle) are carried out by specific enzymes from lignocellulolytic organisms (especially wood-degrading fungi and bacteria). As potential industrial catalysts for biomass conversion, enzymes might provide high specificity, low energy or chemical consumption, or low environment pollution.

Lignocellulosic biomass consists of morphologically different cellulose, structurally and compositionally complex hemicellulose, recalcitrant lignin, diverse proteins, different lipids, and other substances that interact with each other. The primary role of biomass-converting enzymes is to degrade polymeric cellulose or hemicellulose into simple saccharides, sugars which can then be fermented by microorganisms to, or serve as platform molecules for synthesis of valuable fuel or chemicals. Cellulases and hemicellulases can degrade cellulose and hemicellulose to constituent hexoses and pentoses. In general, biomass-converting enzymes have to work in concert, to benefit from synergism among their specificity (towards different components and regions of lignocellulose) as well as mitigation of their inhibition (by different lignocellulose components or degradation products).

Biomass-utilizing organisms, widely distributed in archaea, bacteria, fungi, protists, plants, and animals (including symbiotic gastrointestinal microbes), possess numerous lignocellulolytic enzymes acting on the (hemi) cellulose backbone, hemicellulose substituents, or cellulose-shielding lignin. Many of these enzymes are secreted, either alone or forming supramolecular cellulosome, thus making them promising industrial biocatalysts for biomass conversion. In-depth and systematic basic studies on biomass-active enzymes have been made for decades, and comprehensive reviews for the field have been written (for recent reviews, see [12,13,14,15,16,17,18,19,20,21,22]). In this review, only the most recent developments, especially those relevant to the enzymological aspects of the commercial enzymatic biomass conversion biotechnology, are introduced.

# OVERVIEW

The myriad of biomass-active, lignocellulose-degrading enzymes may be classified in ways emphasizing catalyzed reaction (specificity), structural/evolutionary relation, or other aspects. Based on the International Union of Biochemistry and Molecular Biology's Enzyme Nomenclature and Classification (http://www.chem.qmul.ac.uk/iubmb/enzyme/ [23]), these enzymes belong to EC 3.2.1 glycosidase, EC 4.2.2 lyases, EC 3.1.1 esterases, EC 1.11.1 peroxidases, EC 1.1.3

carbohydrate oxidases, EC 1.10.3 phenol oxidase, and other EC classes, according to their main reactions. Each class and subclass has shared primary enzyme substrates, a feature that may facilitate enzyme selections for targeted biomass materials.

Based on Carbohydrate-Active EnZYmes (http://www.cazy.org/ [24]) and Fungal Oxidative Lignin Enzymes (FOLy) (http://foly.esil.univ-mrs.fr/ [25]) databases, lignocellulose-degrading enzymes belong to Glycoside Hydrolases (GH), Polysaccharide Lyases (PL), Carbohydrate Esterases (CE), Lignin Oxidases (LO), and Lignin Degrading Auxiliary enzymes (LDA) families according to their sequence and structural homology. Each family has shared three-dimensional structure and catalytic mechanism, a feature that may facilitate bioinformatics analysis of (Meta) genomic data. Yet enzymes from different families may catalyze the same reaction.

A distinct structural feature of lignocellulose-degrading enzymes is their modularity. In addition to the catalytic core, many of these enzymes also possess non-catalytic but functionally important domains, including carbohydrate-binding modules (CBM), fibronectin 3-like modules, dockerins, immunoglobulin-like domains, or functionally unknown "X" domains. Having affinity to bundled or individual polysaccharide chains or to single carbohydrate molecules, CBM anchors or directs host enzymes to targeted carbohydrate substrates [26], and in some cases even disrupts crystalline cellulose microfibrils to assist cellulase reaction enzymes [13,27]. Through specific affinity to cohesion, dockerin anchors host enzymes onto scaffoldin to assemble a cellulosome comprising a clustering of different but synergistic/interdependent enzymes [28, 29, and 30]. Modularity equips lignocellulose-degrading enzymes with vast versatility.

Many lignocellulose-degrading enzymes employ hydrolytic reactions (mainly acting on (hemi) cellulose), while others employ oxidoreductive ones (mainly acting on lignin), to convert lignocellulose. Almost all cellulases and hemicellulases are carbohydrate hydrolases relying on either a "retaining" mechanism, which yields product of the same anomeric configuration after breaking a glycosidic bond with a "double-displacement" hydrolysis, or an "inverting" mechanism, which yields product of the opposite anomeric configuration after breaking a glycosidic bond with a "single nucleophilic-displacement" hydrolysis, both involving two acidic amino acid residues (Glu or Asp) as a proton donor or general acid and as a nucleophile or base [31]. "Retaining" hydrolases might also act as glycosyl transferase. All lignin-active peroxidases are heme-containing, some with manganese co-active center, and phenol oxidases are copper-containing oxidoreductases, relying on electron-transfer from lignin to high valence Fe(V/VI)-oxo, Mn(III), or Cu(II), which leads to subsequent radicalization, bond scission, or derivatization in lignin [32].

# Cellulases

Hydrolytic scission of the  (1→4) glucosidic bond in cellulose, leading to the formation of glucose (Glc) and short cellodextrins, is carried out mainly by cellulases, a group of enzymes comprising cellobiohydrolase (CBH), endo-1, 4- -D-glucanase (EG), and  -glucosidase (BG). Although cellulose is relatively simple in terms of composition (anhydro-Glc units only) and morphology (mainly amorphous and monoclinic I or triclinic I crystalline), there is a vast natural diversity of cellulases with catalytic modules belonging to ~14 GH families to accommodate four major reactions modes and different synergisms (for recently studied examples, see [33,34,35]).

# Cellobiohydrolase

Degradation of crystalline cellulose is carried out mainly by CBHs, thus the enzymes are indispensable for industrial enzymatic lignocellulose degradation. Archetypical CBHs are found in GH6 and 7, as well as 48, families. GH7 CBH is found in all known cellulolytic fungi (based on secretome or genome information). GH6 CBH is also found in many cellulolytic fungi. Among secreted proteins and enzymes of cellulolytic fungi, up to 70% wt or so may be CBHs [36, 37, and 38]. Also known as CBH-I (EC 3.2.1.-), GH7 CBH has specificity towards the reducing end of a cellulose chain. In contrast, GH6 CBH, also known as CBH-II (EC 3.2.1.91), can be specific towards the non-reducing end of a cellulose chain. Such "opposing" specificities render GH7 and 6 CBHs highly synergistic and cooperative in degrading their common substrate.

The CBH catalytic core features tunnel-like active sites, a topology that equips CBH with the ability to hydrolyze cellulose "processively": it threads into the end of a cellulose chain through its active site, cleaves off a cellobiosyl unit, glides down the chain, and starts the next hydrolysis step [31, 39]. A CBM may assist the catalytic core with processivity [40]. Such processive reactions, plus the insolubility of the cellulose substrate, makes CBH kinetics deviant from the Michaelis-Menten model, and show significant fractal and "local jamming" effect [41, 42, and 43]. Processive CBH movement can be obstructed by kinks or other impediments on the cellulose surface; and as such it has been suggested that $k_{(off)}$ values may be a major factor in CBH efficiency [44, 45].

GH7 CBH-I may have approximately ten anhydro-Glc-binding subsites within its active tunnel, in which a cellulose segment or cellodextrin is bound and activated via H-bonding and $\varpi$-stacking with key amino acid residues. In addition to the catalytic core, many CBHs also have CBMs, which is believed key in CBH's action on crystalline cellulose.

## Endo-1, 4-β-Glucanase

Degradation of amorphous cellulose can be carried out by EGs (EC 3.2.1.4). Unlike CBH, EG hydrolyzes internal glycosidic bonds in cellulose with a random, on-off fashion. Such dynamics make EG well-suited to less orderly or partially shielded cellulose parts, generating new cellulose chain ends for CBH action. A few EGs can act "processively" on crystalline cellulose [13, 46]. There is a significant synergism between CBH and EG, and their co-presence and cooperation are determinant for highly efficient enzymatic systems of industrial biomass-conversion.

Widely distributed among various organisms, different EGs have a catalytic core belonging to more than ten GH families, of which GH5, 7, 9, 12, 45, and 48 are representative. Typical cellulolytic fungi secrete EGs at ~20% wt level in their secretes [36, 37, 38]. Also known as EG-I, II, III, and V, respectively, GH7, 5, 12, and 45 EG are most common in natural fungal cellulose mixes. Most cellulolytic fungi and bacteria produce numerous EGs. Although they all act on the same cellulose substrate, they do so through differing mechanisms ("inverting" for GH6, 9, 45, and 48 EGs; "retaining" for GH5, 7, 12 EGs). Such EG "plurality" may relate to different EGs' side-activities on hemicellulose in degrading complex lignocellulose [47], or synergism between processive and conventional EGs [13].

The active sites of most EGs are cleft- or groove-shaped, inside which a cellodextrin or a cellulose segment may be bound and acted on by EG. In addition to the catalytic core, EGs may possess CBMs or other domains. CBMs may direct host EG, but is not a pre-requisite, for EG's action.

## B-*Glucosidases*

Degradation of cellobiose, as well as other cellodextrins, is carried out by BG or cellobiose hydrolase (EC 3.2.1.21). Unlike CBH and EG, BGs in general are not modular (lacking distinct CBMs), and have pocket-shaped active sites to act on the non-reducing Glc unit from cellobiose or cello dextrin [48]. BGs belong to the GH1, 3, and 9 families, with GH1 and 3 BGs being archetypical [49]. Unlike the majority of biomass degrading enzymes, the activity of BG, which acts upon soluble rather than insoluble substrate, can be studied using traditional kinetic models [50].

Many cellulolytic fungi produce one or more BGs at levels of about 1% of total secreted proteins, significantly lower than that of CBH and EG [36, 37, 38]. However, BG plays a key role in the efficiency of an enzymatic lignocellulose-degrading system, because its action on cellobiose mitigates product inhibition on CBH and EG. For industrial biomass conversion targeting high feedstock loads, supplementing BG to common microbial cellulolytic enzyme preparations can be imperative, because of high cellobiose level during the enzymatic conversion [51].

GH1 BGs tend to be more resilient to Glc (product) inhibition, as well as more active on different di- or oligosaccharides, than GH3 BGs do. Thus having GH1 BG might enable a cellulolytic enzyme system to be more potent in degrading complex lignocellulose.

# Hemicellulases

In plant cell walls, cellulose is entangled with and shielded by hemicellulose, a group of complex polysaccharides made by different glyco-units and glycosidic bonds. Degradation of hemicellulose, which not only "liberates" cellulose for cellulases but also converts hemicellulose into valuable saccharides, is carried out mainly by an array of interdependent and synergistic hemicellulases. Common hemicelluloses include -glucan, xylan, xyloglucan, arabinoxylan, mannan, galactomannan, arabinan, galactan, polygalacturonan, etc., which are targets of -glucanase, xylanase, xyloglucanase, mannanase, arabinase, galactanase, polygalacturonase, glucuronidase, acetyl xylan esterase, and other enzymes [22, 52]. Among hemicellulases, glycoside hydrolases (belonging to about 29 GH families) hydrolyze glycosidic bonds, carbohydrate esterases (belonging to about 9 CE families) hydrolyze ester bonds, and polysaccharide lyases (belonging to about 5 PL families) cleave glycosidic bonds. endo-Hemicellulases cleave internal/backbone glycosidic bonds, whereas other glycosidases remove mainly the chain's substituents or side chains. Cellulolytic microbes produce many hemicellulases along with cellulases for effective lignocellulose degradation (for recently studied examples see [36, 37, and 38]).

Different plants have different hemicelluloses: acetylated (galacto) glucomannan (as well as arabinoglucuronoxylan), glucuronoxylan, and arabinoxylan are major hemicellulose in softwood, hardwood, and grass, respectively [52]. Hence different hemicellulase combinations are needed for different biomass feedstocks in industrial biomass conversion. Synergism of hemicellulases is found both amongst hemicellulases themselves and between hemicellulases and cellulases [2, 35, 53, 54, 55, and 56].

## Endo-β-Xylanases and β-Xylosidase

Degradation of (glucurono)(arabino)xylan, a group of (1→4) linked D-xylopyranosyl (Xyl) polysaccharides with different O-substitutions by acetyl, glucuronoyl (GlcU), arabinosyl (Ara), or other substituents, is mainly carried out by endo-xylanase (EX, EC 3.2.1.8), which hydrolyzes backbone glycosidic bonds in xylan. Widely distributed among archaea, bacteria, fungi, and plants, EXs have catalytic cores belonging to the GH8, 10, 11, 30, and 43 families, with GH10 and GH11 EX being archetypical [57]. GH10 and 11 EX differ in substrate specificity: GH10 EX produces shorter oligosaccharides and has more activity on substituted xylan [58]. Besides the catalytic core, one or more CBMs or other domains may be found in EXs [59]. As BG does for EG, -xylosidases (BX, EC 3.2.1.37) hydrolyze xylobiose or other xylooligosaccharides, after their production from xylan by EX [60].

BXs have catalytic cores belonging to the GH3, 30, 39, 43, 52, and 54 families. Many BXs have -L-arabinofuranosidase activity. Like cellulases, xylanases also employ either an "inverting" or a "retaining" mechanism based on a nucleophile and a general acid catalytic diad for their catalysis.

Among enzymes secreted by cellulolytic fungi, xylanases often account for <1% wt, although multiple xylanases may be produced [36, 37, 38]. Xylanases may provide, or benefit from, significant synergism among themselves, with other hemicellulases, or with cellulases: GH11 EX may produce large xylooligosaccharides for GH10 EX action, debranching hemicellulases may remove substituents to facilitate EX reaction, or EX may degrade xylan in lignocellulose to expose cellulose for cellulase reaction [61]. Hence, optimizing a lignocellulose-degrading enzyme mix with xylanases can be key for industrial biomass conversion, especially for hardwood or grass-based feedstocks enriched in (arabion) xylan.

## Acetyl Xylan Esterase, Feruloyl Esterase, and Glucuronoyl Esterase

Acetyl, feruloyl (or other hydroxycinnamoyls), and GlcU groups are common ester substituents in xylan or other hemicelluloses. Their removal, often key for effective EX or other endo-hemicellulases activity, is carried out by acetyl xylan esterase (AXE, EC 3.1.1.72), feruloyl esterase (FAE, EC 3.1.1.73), and glucuronoyl esterase (GE, EC 3.1.1.-), respectively. Belonging to CE1, 2, 3, 4, 5, 6, 7, and 12 families, AXE deacetylates substituted O2 or O3 sites of backbone glycosyl units in xylan or other hemicelluloses [62]. Belonging to CE1 family, FAE hydrolyzes feruloyl esters at -L-Ara (O2 or O5 site), -D-galactosyl (Gal, O6 site), or -D-Xyl side chains of arabinan/arabinoxylan, rhamnogalacturonan, or xyloglucan [63]. Belonging to CE15 family, GE demethylates O6-methyl glucuronoyl (GlcU) $(1{\rightarrow}2)$ linked to backbone Xyl in glucuronoarabinoxylan [64]. These esterases have the canonical Ser-His-Asp catalytic triad found in other esterases, lipases, or serine proteases.

FAEs may have CBMs together with their catalytic core. Different FAEs have different specificity towards different hydroxycinnamoyl ester bonds, which are involved in linking hemicellulose to lignin

[65]. Different AXEs, FAEs, and GEs may cooperate in attacking complex hemicellulose. AXE or FAE can assist endo-hemicellulases by deacetylating, deferulating, or delignifying hemicelluloses. GEs can assist -glucuronidase by hydrolyzing GlcU ester. For industrial biomass conversion, supplementing AXE or FAE may enhance endo-hemicellulases' activity, by attacking acetylated hardwood xylan or ferulated grass arabinoxylan.

## α-L-Arabinofuranosidase, α -Galactosidase and α -Glucuronidase

Removal of Ara substituent is carried out by -L-arabinofuranosidase (AF, EC 3.2.1.55), a group of enzymes whose catalytic cores belong to GH3, and 43, 51, 54, and 62 families [66]. Many AFs contain CBMs. There are specificity differences among AFs: some prefer single Ara esterifying either O2, 3, or 5 site, while others prefer dual Ara esterifying O2 and 3 sites.

Removal of Gal substituent linked via -glycosidic bonds to galactomannan, pectin, or other hemicelluloses is carried out by -galactosidase (EC 3.2.1.22), a group of enzymes whose catalytic cores belong to GH4, and 27, 36, 57, and 110 families. GH4 and 110 -galactosidases rely on $NAD^+$ cofactor and a redox mechanism to hydrolyze their substrates, unique among known glycosidases [67].

Removal of $(1\rightarrow2)$ linked glucuronoyl or its methyl ester in xylan (often at the O2 site) or other hemicelluloses is carried out by -glucuronidases (AG, EC 3.2.1.139), a group of enzymes whose catalytic cores belong to GH67 and 115 families [68, 69]. Some AGs have higher specificity to glucuronated xylooligosaccharides, while others have higher specificity to polymeric glucuronoxylan [61].

AF, -galactosidase, and -glucuronidases may assist xylanase, pectinase, or other hemicellulases by debranching their polymeric substrates. Supplementing AF to lignocellulose-degrading enzyme mix may be highly beneficial when softwood (abundant in arabinoglucuronoxylan) or grass (rich in arabinoxylan) feedstocks are targeted.

# Glucanase, Mannanase, Xyloglucan Hydrolase and Pectinase

Degradation of (1→3), (1→4), or (1→6) glucan can be carried out by (non-EG or BG) -glucanases (EC 3.2.1.-), a diverse group of endo- or exo-acting, glycosidic bond type-specific or promiscuous enzymes whose catalytic cores belong to GH3, 5, 12, 16, 17, 55, 64, and 81 families [70]. Many cellulolytic microbes secrete, or anchor in cell membrane, one or more -glucanases, whose differential specificities may enable cooperative degradation of complex -glucans heterogeneous in backbone architecture or glycosidic bonds.

Degradation of (galacto)(gluco)mannans, (1→4)-D-mannosyl or manno/glucopyranosyl polymers with variable (1→6) D-Gal side chain as well as O2 and/or O3 acetylation, can be carried out by mannanase (EC 3.2.1.78), a group of widely distributed, hydrolytic enzymes with catalytic cores belonging to GH5, 26, and 113 families [71]. Mannooligosaccharides produced by mannanase can be further degraded by -mannosidases (EC 3.2.1.25), whose catalytic cores belong to GH1, 2, and 5 families. In addition to a catalytic core, mannanases may possess one or more CBM (specific to mannan or cellulose) or other domains (e.g., Ig-like and S-layer module). Many cellulolytic microbes co-secrete mannanases with cellulases, xylanases, and other enzymes [36]. Feedstocks abundant in (galacto)glucomannan, such as softwood, are likely in need of enzyme mixes containing sufficient mannanase activity for effective substrate degradation/conversion.

Degradation of xyloglucan, (1→4) glucan with (1→6) linked Xyl substituted by either (1→2) L-Ara or (1→2) D-Gal units (partially acetylated or substituted by (1→2) L-fucopyranosyl (Fuc)), can be carried out by xyloglucan hydrolases (EC 3.2.1.150, 151, 155), whose catalytic cores belong to GH5, 12, 16, 44, and 74 families, and that are part of the xyloglucan transferase/hydrolase (XTH) superfamily [72]. Many xyloglucan hydrolases have minor EG activity, while many EGs have minor xyloglucan-hydrolyzing activity [47]. AFs and esterases may de-branch xyloglucan, thus allowing more effective xyloglucan hydrolase activity. Xyloglucan hydrolases may be important for industrial biomass conversion, because the degradation of xyloglucan could enhance cellulase accessibility to cellulose [73]. A diverse group of pectinolytic enzymes are responsible for the degradation of pectic polysaccharides, consisted of (1→4) poly- -(rhamno)galacturonic acids with variable backbone methylation/acetylation and Ara and Gal side chains branching [74,75]. Common pectinolytic enzymes include polygalacturonases (EC 3.2.1.15, 67, 82) with catalytic cores belonging to GH28 family, pectin/pectate lyases (EC 4.2.2.2, 6, 9, 10) with catalytic cores belonging to PL1, 2, 3, 9, and 10 families, and pectin methyl esterase (EC 3.1.1.11) with catalytic cores belonging to CE8 family. Unlike hydrolytic polygalacturonases, pectin/pectate lyases cleave an O-C4 glycosidic bond, assisted by C6-uronate, to form a Δ4:5 C=C bond at the non-reducing side of galacturonoyl unit. Many pectinolytic hydrolases and lyases act on both pectin and pectate. Different pectinolytic enzymes act in concert: endo- and exo-enzymes synergize each other as EG and CBH do; hydrolases and lyases may act on different parts of pectin/pectate; and methyl esterases demethylate pectin to help pectate-specific enzymes. AF, galactosidase, and other enzymes may enhance pectinolytic enzymes' action by removing the side chains from polyrhamnogalacturonan. Rich in pectic polysaccharides, sugar beet pulp and fruit residue-based feedstocks are likely in need of sufficient pectinolytic enzymes for effective enzymatic degradation/conversion.

## Lignocellulose Oxidoreductases

The secretomes of most cellulolytic microbes (particularly white and brown rots) contain oxidoreductases, in some cases at quite significant levels, whose co-presence with hydrolytic or lytic enzymes indicates the importance of having an oxidoreductive system as part of effective biological lignocellulose degradation [5,32,37,76,77]. The main task of these oxidoreductases is likely aimed at degradation of lignin, a highly heterogeneous and recalcitrant aromatic polymer (consisted of various syringyl, guaiacyl, or other hydroxyphenyl units) entangled with hemicellulose or cellulose and inactive/inhibitory to (hemi) cellulases. Lignin degradation is imperative for industrial enzymatic biomass-conversion, because it not only increases (hemi) cellulose accessibility for (hemi)cellulase but also diminishes (hemi)cellulase inactivation caused by lignin adsorption.

Lignin peroxidase (EC 1.11.1.14), Mn peroxidase (EC 1.11.1.13), and versatile peroxidase (EC 1.11.1.16) are extracellular fungal heme peroxidases (belonging to LO2 family) with high potency to oxidatively degrade lignin. Upon interaction with $H_2O_2$, these enzymes form highly reactive Fe(V) or Fe(IV)-oxo species, which abstract electrons from lignin (to cause oxidation or radicalization) either directly or via Mn(III) species. Laccase (EC 1.10.3.2) is a multi-copper oxidase (belonging to LO1 family) secreted by numerous lignocellulolytic fungi. This enzyme can directly oxidize phenolic parts of lignin, or indirectly oxidize non-phenolic lignin parts with the aid of suitable redox-active mediator.

Aryl-alcohol oxidase (EC 1.1.3.7), glyoxal oxidase (EC 1.1.3.-), and various carbohydrate oxidases (EC 1.1.3.4, 9, 10) are also involved in natural lignocellulose degradation. These enzymes, belonging to LDA1–6 families, can generate $H_2O_2$ from $O_2$, with concomitant oxidation of aromatic alcohol, glyoxal, or reducing carbohydrates. The generated $H_2O_2$ may support lignin-degrading peroxidases or power Fenten-type chemistry that degrades lignin non-enzymatically.

Cellobiose dehydrogenase (EC 1.1.99.18) is produced by many lignocellulotic fungi. The flavoheme enzyme (belonging to LO3 family) may dehydrogenate or oxidize cellobiose or other cellodextrins to corresponding aldonolactones, with concomitant quinone reduction to phenol or $O_2$ reduction to $H_2O_2$. Until recently, the role of CDH in enzymatic degradation of lignocellulose was largely considered only from a Fenton chemistry perspective [78] or mitigation of cellulases' product inhibition by cellobiose [79]; however, the recent discovery of the CDH stimulation of GH61 enzyme activity may shed new light upon its function [80, 81, 82] (see Section 3.1).

The lignin-degrading oxidative species generated by these oxidoreductases may attack inhibitors of industrial sugar-to-fuel/chemical microbial conversion, making the process more effective [83]. However, they may also attack (hemi) cellulose or even (hemi) cellulases, potentially hampering industrial enzymatic (hemi)cellulose conversion [84,85]. For the peroxidases, autooxidation or inactivation may severely limit their performance. Applying such enzymes to industrial enzymatic biomass conversion needs further research for overall benefit.

# EMERGENT INDUSTRIAL LIGNOCEL-LULOSE-DEGRADING ENZYMES

In addition to the enzyme categories above, lignocellulose degrading microbes encode and secrete a number of proteins, often co-induced and secreted with canonical (hemi) cellulases, with more enigmatic functionalities. Some of these molecules, when added to cellulases, result in large increases in lignocellulose breakdown, while others display less dramatic stimulatory effects on cellullase activity (at least under laboratory conditions). An improved understanding to how these enzymes function may result in significant improvements in industrial lignocellulose degradation, and as such, much attention is currently focused on these molecules.

## Discovery of GH61 Cellulase-Enhancing Protein

Many cellulolytic fungi encode and express proteins which are classified as members of the Glycoside Hydrolase 61 (GH61) family [36, 37, 38], but due to the unusual characteristics of the GH61s, it took more than a decade after the initial discovery of this group of enzymes to begin to understand their function. The first of these proteins was found during a cDNA screen for the Trichoderma reesei complement to the Saccharomyces cerevisiae gene SEC1, which is involved in secretion. Unexpectedly, several of the isolated T. reesei cDNA clones, which were shown to suppress the temperature sensitive sec1-1 mutant of S. cerevisiae, contained a consensus sequence for a CBM. Based on this CBM homology, the protein encoded by this cDNA was expressed and tested for cellulase activity. The prepared protein sample was observed to have weak activity in the breakdown of -glucan, carboxymethyl cellulose (CMC), and phosphoric acid-swollen cellulose (PASC), and was classified as T. reesei EG-IV [86]. A further characterization of this protein reported that this enzyme could cause the release of small amounts of soluble cellooligosaccharides from cellulosic substrates [87]. Similar results, showing weak if any EG activity, were reported for a homologous GH61 encoded by Aspergillus kawachii [88].

Suspicion that GH61 was not an authentic EG got strong support from GH61 crystal structures. EGs usually have a cleft in which the cellulose strand binds to the enzyme, typically through interactions with aromatic residues, and a hydrolase active site near this cleft containing a catalytic glutamic or aspartic acid. The structure of T. reesei GH61B [89] contains neither a cleft nor an obvious hydrolase active site, but rather displays a surface-exposed divalent metal binding site surrounded by hydrophobic residues, suggesting a possible carbohydrate binding surface and active site of unknown function. Structurally, T. reesei GH61B is homologous to chitin binding protein 21 (CBP21, belonging to CBM33 family) from the bacterium Serratia marcescens, a molecule which at the time was thought to be a non-catalytic protein which enhanced chitin breakdown by chitin hydrolases [90]. Another publication reported the structure of Thielevia terrestris GH61E, which also includes a surface exposed metal binding site; this report also included significant biochemical analysis of T. terrestris GH61E, T. terrestris GH61B, and Thermoascus aurantiocus GH61A activity upon both model and industrial cellulosic substrates [91]. In this analysis, the GH61 molecules tested displayed only negligible ability to cleave any of the cellulosic or hemicellulosic substrates tested. However, addition of GH61s to Trichoderma cellulase cocktails was shown to greatly enhance the activity of these cellulases in lignocellulose degradation, lowering the required enzyme concentration for substrate breakdown by a factor of two. Confirming the role of metal in GH61 activity, the GH61 cellulase-boosting effect was inhibited by addition of EDTA, and metal binding site mutants of GH61 displayed greatly reduced or absent cellulase boosting activity. Surprisingly, the addition of GH61 has no stimulatory effect on Trichoderma cellulases' breakdown of "pure" cellulose substrates (e.g., microcrystalline cellulose) suggesting that a non-cellulose component might be a requirement for or the target of GH61 activity [91]. While the industrial significance of GH61 as a critical component in cellulase mixtures was established, the biochemical activity of these molecules remained mysterious.

Recently, the functions of members of the fungal GH61 family, and members of the related bacterial CBM33 family, were revealed. Several of these enzymes have been shown to act by cleaving cellulose [80,81,82,92,93,94,95,82,92] or the structurally related polysaccharide chitin [96]; and while some differences in data and analysis exist amongst these publications it is clear that GH61 and CBM33 enzymes do not function as typical glycoside hydrolases.

Unlike canonical glycoside hydrolases, GH61 and CBM33 enzymes require a redox-active factor for activity. This cofactor requirement explains the formerly puzzling results which show little or no GH61 activity upon model cellulose substrates, but a substantial increase in complex biomass degradation by Trichoderma cellulases or mixtures upon GH61 supplementation [91]; when the soluble phase of dilute acid pretreated biomass was added to model cellulose substrates, GH61 was shown to act upon those substrates, indicating that requisite GH61 cofactor was present in soluble (and perhaps also in insoluble) fractions of pretreated biomass [94]. GH61 and CBM33 polysaccharide cleaving-activity has been demonstrated upon addition of a small molecule reducing-agent such as ascorbate, glutathione, or gallate [92, 93, 94, 95, 96] or the fungal enzyme cellobiose dehydrogenase [80, 81, 82]. Analysis of the polysaccharide cleavage products formed by GH61 and CBM33 molecules reveals that these enzymes release oxidized cellooligosaccharides, though significant quantities of non-oxidized oligosaccharides are also detected in some studies [93, 94, and 95]. Unlike other glycoside hydrolases, which are more active on cellooligosaccharides than crystalline cellulose, GH61 and CBM33 enzymes appear inactive upon cellooligosaccahrides [81, 93, 94, 95, and 96]. The position of the oxidation on the oligosaccharide products has been reported on the reducing end [81, 82, 93, 95], non-reducing end [81, 82], or both [94], which could suggest differences amongst these enzymes. GH61 proteins also require a metal for this cellulose cleaving activity, specifically copper which binds tightly to the protein in a type-2 copper site geometry [82, 94, 96]. Interestingly, despite the large potential of this enzyme class to promote lignocellulose breakdown, the precise cleavage mechanism of GH61 and CBM33 enzymes is unclear, and is the subject of ongoing study by multiple groups.

## Expansin, Swollenin, and Loosinin

Expansins are a class of plant proteins which interact with and modify cell walls and/or cell wall components by an unknown activity, thought to result in expansion, slippage, or lengthening of cell wall structures. These two-domain proteins consist of a domain homologous to the GH45 EG catalytic core and a second domain homologous to Group II grass pollen allergens; both domains have no known catalytic function and display no detectable hydrolytic activity on lignocellulosic or model substrates (expansins are thoroughly reviewed by [97]). T. reesei has been shown to express a protein, named swollenin, which has sequence homology to plant expansins and displays a similar mysterious disruptive effect on cellulosic substrates. Isolated T. reesei swollenin has been shown, without cause to formation of detectable reducing ends, to weaken filter paper and to affect superstructural changes in cotton fibrils by light and atomic force microscopy [98].

Despite the apparent lack of direct lytic activity on cellulose, the addition of T. reesei swollenin significantly increases the breakdown of filter paper by cellulases [99,100]. Swollenin from Aspergillus fumigatus is reported to have weak lytic activity on CMC but no apparent hydrolytic activity on microcrystalline cellulose, though treatment of microcrystalline cellulose with A. fumigatus swollenin reduced apparent microcrystalline cellulose particle size and potentiated breakdown of the cellulose by hydrolytic cellulases [101]. The basidiomycete Bjerkandera adusta produces a similar protein, loosinin, which increases cellulase activity on cotton and agave bagasse [102]. Bacterial species, including Bacillus subtilis [103] and Hahella chejuensis [104], also produce expansin-like molecules. Like fungal swollenins, bacterial expansins alter cellulose fiber structure and promote the breakdown of cellulose by hydrolases without showing detectable direct hydrolase activity [104,105]. Recent structural and mutational studies of expansin EXLX1 from B. substilis have shed some light on these proteins, demonstrating that several clustered residues on GH45-like domain are required for EXLX1's cell wall modifying activity, and that the second domain is likely a new type of CBM [103,106]. Further study will be required in order to determine the mechanism by which these molecules increase lignocellulose conversion and allow industrial exploitation of this class of protein.

# Cip Proteins

CIP1 and CIP2 (cellulose induced protein-1 and -2, respectively) were first found in a transcriptional analysis of T. reesei. Both contain a CBM and are co-regulated with known cellulases [107]. The function of CIP1 is unknown, though it is claimed that CIP1 from T. reesei has weak activity on p-nitrophenyl -D-cellobioside [108] and some synergistic activity with both GH61 and swollenin [109]. CIP2, found in both T.

reesei and Schizophyllum commune, has recently been shown to be an esterase that cleaves the methyl ester of 4-O-methyl-D-glucuronic acid [110]. This enzyme, now classified as the first member of CE15 family (EC 3.1.1.-), likely acts in the cleavage of hemicellulose-lignin crosslinks. Further investigation of both the functions and the potential of these enzymes in industrial applications are needed.

## Cellulosomes

While many microorganisms secrete biomass degrading enzymes into their environment, other microbes, particularly anaerobic biomass degrading microbes, use cell-surface linked enzymes to break down lignocellulosic materials. Cellulosomes are arrays of multiple cellulase and hemicellulase proteins, assembled by specific interactions between dockerin domains on the enzyme and cohesins bound to structural scaffoldins on the microbial surface (reviewed extensively in [28,29,30,111]). This spatial clustering of multiple lignocellulose degrading enzymes results in an increased synergy between lytic activities [28, 30, 112]. It has been shown that recombinant cellulosomes can be transplanted to other industrially useful organisms, such as S. cerevisiae [113,114] and B. subtilis [115]. The ability of cellulosomes to cluster activities may present unique capabilities, both in synergistic breakdown of a substrate and in targeted degradation of specific biomass components.

# PERSPECTIVES

Microbial breakdown of lignocellulosic biomass is a highly complex process, requiring multiple types of synergistic catalytic activities simultaneously acting upon a variety of both soluble and insoluble polymeric substrates. Industrial biomass utilization processes will also require a combination of activities in order to degrade lignocellulosic feedstocks, but with the additional complication that thermal and chemical pretreatment processes, which improve physical substrate access to the enzymes, alter the substrate chemically and form or solubilize inhibitory compounds that can negatively impact enzyme performance [116]. When combined with traditional cellulases, previously uncharacterized enzymes, some of which have been discussed in this review, have a high degree of synergy in the degradation of pretreated lignocellulosic substrates [91]. Further improvements in and synergies with industrial processes are possible with other emergent enzyme families, and with yet to be discovered or currently overlooked enzyme activities. The exploration of biological diversity, particularly from extremophilic organisms, may reveal new cellulases with improved properties for specific industrial processes. For example, it has been shown that cellulases from halophilic organisms have higher resistance to inactivating ionic liquid residues from certain pretreatments processes [117,118]. Protein engineering of enzymes, such as components of the T. reesei cellulase complex, to improve their suitability for industrial biomass applications has been a growing area of activity within both academic and industrial research into biomass conversion. Many groups report improvements to the properties of biomass degrading enzymes, such as increased thermostability or thermoactivity, altered pH optima, decreased lignin binding or glucose inhibition, or improved activity on crystalline cellulose.

This review has focused on specific enzymes and activities. However, the production of these enzymes, or more likely mixtures of enzymes, must also be considered in order to develop viable enzymatic biomass conversion technologies. An enzyme mixture with improved activity but which is prohibitively expensive to produce on a large scale will be of little industrial use. One topic of great consequence, but which is beyond the scope of this review, is the development of production hosts for biomass degrading enzymes. Significant efforts have been made, by either natural diversity exploration, classical mutagenesis, or genetic engineering, which have resulted in strains with improved enzyme expression profiles, higher protein secretion levels, or the ability to utilize pretreated biomass as a nutrient source. However, the required activities, or ratios of activities, which will be required for optimal breakdown of lignocellulose will vary among different types of pretreatment and biomass, making the ability to design strains, and the enzymes they express, optimized for a specific process or robust for many processes, an area of mounting importance.

While the enzymatic breakdown of lignocellulosic biomass is a complicated process, involving many activities which work in tandem to decompose a heterogeneous and recalcitrant substrate, the understanding of these enzymes and activities has increased significantly in recent years. Ongoing studies of both known and yet-to be discovered enzymes will provide further insight into this complex process and give guidance as to how enzymes can be better applied to a variety of industrial processes.

# ACKNOWLEDGEMENTS

We thank Robert L. Starnes of Novozymes for critical reading of this manuscript.

# REFERENCES

1.  Lynd, L.R.; Laser, M.S.; Bransby, D.; Dale, B.E.; Davison, B.; Hamilton, R.; Himmel, M.; Keller, M.; McMillan, J.D.; Sheehan, J.; Wyman, C.E. How biotech can transform biofuels. Nat. Biotechnol. 2008, 26, 169–172.

2.  Kumar, R.; Singh, S.; Singh, O.V. Bioconversion of lignocellulosic biomass: biochemical and molecular perspectives. J. Ind. Microbiol. Biotechnol. 2008, 35, 377–391.

3.  Wackett, L.P. Biomass to fuels via microbial transformations. Curr. Opin. Chem. Biol. 2008, 12, 187–193.

4.  Margeot, A.; Hahn-Hagerdal, B.; Edlund, M.; Slade, R.; Monot, F. New improvements for lignocellulosic ethanol. Curr. Opin. Biotechnol. 2009, 20, 372–380.

5.  Dashtban, M.; Schraft, H.; Qin, W. Fungal bioconversion of lignocellulosic residues; opportunities & perspectives. Int. J. Biol. Sci. 2009, 5, 578–595.

6.  Sánchez, C. Lignocellulosic residues: biodegradation and bioconversion by fungi. Biotechnol. Adv. 2009, 27, 185–194.

7.  Xu, F. Biomass-converting enzymes and their bioenergy applications. In The Manual of Industrial Microbiology and Biotechnology, 3rd; Baltz, R.H., Demain, A.L., Davies, J.E., Bull, A.T., Junker, B., Katz, L., Lynd, L.R., Masurekar, P., Reeves, C.D., Zhao, H., Eds.; American Society for Microbiology Press: Washington, DC, USA, 2010; pp. 495–508.

8.  Girio, F.M.; Fonseca, C.; Carvalheiro, F.; Duarte, L.C.; Marques,

S.; Bogel-Lukasik, R. Hemicelluloses for fuel ethanol: A review. Bioresour. Technol. 2010, 101, 4775–4800.

9.  Sims, R.E.H.; Mabee, W.; Saddler, J.N.; Taylor, M. An overview of second generation biofuel technologies. Bioresour. Technol. 2010, 101, 1570–1580.

10. Chandel, A.K.; Singh, O.V. Weedy lignocellulosic feedstock and microbial metabolic engineering: advancing the generation of "Biofuel". Appl. Microbiol. Biotechnol. 2011, 89, 1289–1303.

11. Zhang, Y.H.P. What is vital (and not vital) to advance economically-competitive biofuels production. Process Biochem. 2011, 46, 2091–2110.

12. Henrissat, B.; Sulzenbacher, G.; Bourne, Y. Glycosyltransferases, glycoside hydrolases: surprise, surprise! Curr. Opin. Struct. Biol. 2008, 18, 527–533.

13. Wilson, D.B. Three microbial strategies for plant cell wall degradation. Ann. N. Y. Acad. Sci. 2008, 1125, 289–297.

14. Wilson, D.B. Cellulases and biofuels. Curr. Opin. Biotechnol. 2009, 20, 295–299.

15. Baldrian, P.; Valásková, V. Degradation of cellulose by basidiomycetous fungi. FEMS Microbiol. Rev. 2008, 32, 501–521.

16. Blumer-Schuette, S.E.; Kataeva, I.; Westpheling, J.; Adams, M.W.; Kelly, R.M. Extremely thermophilic microorganisms for biomass conversion: status and prospects. Curr. Opin. Biotechnol. 2008, 19, 210–217.

17. Gowen, C.M.; Fong, S.S. Exploring biodiversity for cellulosic biofuel production. Chem. Biodivers. 2010, 7, 1086–1097.

18. Jovanovic, I.; Magnuson, J.K.; Collart, F.; Robbertse, B.; Adney, W.S.; Himmel, M.E.; Baker, S.E. Fungal glycoside hydrolases for saccharification of lignocellulose: outlook for new discoveries fueled by genomics and functional studies. Cellulose 2009, 16, 687–697.

19. Yeoman, C.J.; Han, Y.; Dodd, D.; Schroeder, C.M.; Mackie, R.I.; Cann, I.K.O. Thermostable enzymes as biocatalysts in the biofuel industry. Adv. Appl. Microbiol. 2010, 70, 1–55.

20. De Vries, R.P.; Battaglia, E.; Coutinho, P.M.; Henrissat, B.; Visser,

J. (Hemi-)cellulose degrading enzymes and their encoding genes from Aspergillus and Trichoderma. In The Mycota X. Industrial Applications, 2nd; Hofrichter, M., Ed.; Springer-Verlag: Berlin, Germany, 2010.

21. Gilbert, H.J. The biochemistry and structural biology of plant cell wall deconstruction. Plant Physiol. 2010, 153, 444–455.

22. Van den Brink, J.; de Vries, R.P. Fungal enzyme sets for plant polysaccharide degradation. Appl. Microbiol. Biotechnol. 2011, 91, 1477–1492.

23. Enzyme Nomenclature; Webb, E.C., Ed.; Academic Press: San Diego, CA, USA, 1992.

24. Cantarel, B.L.; Coutinho, P.M.; Rancurel, C.; Bernard, T.; Lombard, V.; Henrissat, B. The Carbohydrate-Active EnZymes database (CAZy): An expert resource for glycogenomics. Nucleic Acids Res. 2009, 37, D233–D238.

25. Levasseur, A.; Piumi, F.; Coutinho, P.M.; Rancurel, C.; Asther, M.; Delattre, M.; Henrissat, B.; Pontarotti, P.; Asther, M.; Record, E. FOLy: An integrated database for the classification and functional annotation of fungal oxidoreductases potentially involved in the degradation of lignin and related aromatic compounds. Fungal Genet. Biol. 2008, 45, 638–645.

26. Guillen, D.; Sanchez, S.; Rodriguez-Sanoja, R. Carbohydrate-binding domains: Multiplicity of biological roles. Appl. Microbiol. Biotechnol. 2010, 85, 1241–1249.

27. Moser, F.; Irwin, D.; Chen, S.L.; Wilson, D.B. Regulation and characterization of Thermobifida fusca carbohydrate-binding module proteins E7 and E8. Biotechnol. Bioeng. 2008, 100, 1066–1077.

28. Ding, S.Y.; Xu, Q.; Crowley, M.; Zeng, Y.; Nimlos, M.; Lamed, R.; Bayer, E.A.; Himmel, M.E. A biophysical perspective on the cellulosome: new opportunities for biomass conversion. Curr. Opin. Biotechnol. 2008, 19, 218–227.

29. Gilbert, H.J. Cellulosomes: microbial nanomachines that display plasticity in quaternary structure. Mol. Microbiol. 2007, 63, 1568–1576.

30. Bayer, E.A.; Lamed, R.; Himmel, M.E. The potential of cellulases

and cellulosomes for cellulosic waste management. Curr. Opin. Biotechnol. 2007, 18, 237–245.

31. Vocadlo, D.J.; Davies, G.J. Mechanistic insights into glycosidase chemistry. Curr. Opin. Chem. Biol. 2008, 12, 539–555.

32. Martínez, A.T.; Speranza, M.; Ruiz-Dueñas, F.J.; Ferreira, P.; Camarero, S.; Guillén, F.; Martínez, M.J.; Gutiérrez, A.; del Río, J.C. Biodegradation of lignocellulosics: microbial, chemical, and enzymatic aspects of the fungal attack of lignin. Int. Microbiol. 2005, 8, 195–204.

33. Duncan, S.M.; Schilling, J.S. Carbohydrate-hydrolyzing enzyme ratios during fungal degradation of woody and non-woody lignocellulose substrates. Enzym. Microb. Technol. 2010, 47, 363–371.

34. Zhang, M.; Su, R.; Qi, W.; He, Z. Enhanced enzymatic hydrolysis of lignocellulose by optimizing enzyme complexes. Appl. Biochem. Biotechnol. 2010, 160, 1407–1414.

35. Banerjee, G.; Car, S.; Scott-Craig, J.S.; Borrusch, M.S.; Walton, J.D. Rapid optimization of enzyme mixtures for deconstruction of diverse pretreatment/biomass feedstock combinations. Biotechnol. Biofuels 2010, 3, 22.

36. Herpoël-Gimbert, I.; Margeot, A.; Dolla, A.; Jan, G.; Mollé, D.; Lignon, S.; Mathis, H.; Sigoillot, J.-C.; Monot, F.; Asther, M. Comparative secretome analyses of two Trichoderma reesei Rut-C30 and CL847 hypersecretory strains. Biotechnol. Biofuels 2008, 1, 18.

37. Sipos, B.; Benko, Z.; Dienes, D.; Reczey, K.; Viikari, L.; Siika-aho, M. Characterisation of specific activities and hydrolytic properties of cell-wall-degrading enzymes produced by Trichoderma reesei Rut C30 on different carbon sources. Appl. Biochem. Biotechnol. 2010, 161, 347–364.

38. Chundawat, S.P.S.; Lipton, M.S.; Purvine, S.O.; Uppugundla, N.; Gao, D.; Balan, V.; Dale, B.E. Proteomics-based compositional analysis of complex cellulase-hemicellulase mixtures. J. Proteome Res. 2011, 10, 4365–4372.

39. Liu, Y.S.; Baker, J.O.; Zeng, Y.; Himmel, M.E.; Haas, T.; Ding, S.Y. Cellobiohydrolase hydrolyzes crystalline cellulose on hydrophobic faces. J. Biol. Chem. 2011, 286, 11195–11201.

40. Beckham, G.T.; Matthews, J.F.; Bomble, Y.J.; Bu, L.; Adney, W.S.; Himmel, M.E.; Nimlos, M.R.; Crowley, M.F. Identification of amino acids responsible for processivity in a Family 1 carbohydrate-binding module from a fungal cellulase. J. Phys. Chem. B 2010, 114, 1447–1453.

41. Xu, F.; Ding, H. A new kinetic model for heterogeneous (or spatially confined) enzymatic catalysis: Contributions from the fractal and jamming (overcrowding) effects. Appl. Catal. A Gen. 2007, 317, 70–81.

42. Igarashi, K.; Uchihashi, T.; Koivula, A.; Wada, M.; Kimura, S.; Okamoto, T.; Penttilä, M.; Ando, T.; Samejima, M. Traffic jams reduce hydrolytic efficiency of cellulase on cellulose surface. Science 2011, 333, 1279–1282.

43. Warden, A.C.; Little, B.A.; Haritos, V.S. A cellular automaton model of crystalline cellulose hydrolysis by cellulases. Biotechnol. Biofuels 2011, 4, 39.

44. Kurasin, M.; Väljamäe, P. Processivity of cellobiohydrolases is limited by the substrate. J. Biol. Chem. 2011, 286, 169–177.

45. Praestgaard, E.; Elmerdahl, J.; Murphy, L.; Nymand, S.; McFarland, K.C.; Borch, K.; Westh, P. A kinetic model for the burst phase of processive cellulases. FEBS J. 2011, 278, 1547–1560.

46. Li, Y.; Irwin, D.C.; Wilson, D.B. Increased crystalline cellulose activity via combinations of amino acid changes in the family 9 catalytic domain and family 3c cellulose binding module of Thermobifida fusca Cel9A. Appl. Environ. Microbiol. 2010, 76, 2582–2588.

47. Vlasenko, E.; Schülein, M.; Cherry, J.; Xu, F. Substrate specificity of family 5, 6, 7, 9, 12, and 45 endoglucanases. Bioresour. Technol. 2010, 101, 2405–2411.

48. Langston, J.; Sheehy, N.; Xu, F. Substrate specificity of Aspergillus oryzae family 3 -glucosidase. Biochim. Biophys. Acta 2006, 1764, 972–978.

49. Eyzaguirre, J.; Hidalgo, M.; Leschot, A. Beta-glucosidases from filamentous fungi: Properties, structure, and applications. In Handbook of Carbohydrate Engineering; Yarema, K.J., Ed.; CRC Press: Boca Raton, FL, USA, 2005; pp. 645–685.

50. Jeoh, T.; Baker, J.O.; Ali, M.K.; Himmel, M.E.; Adney, W.S.

Beta-D-glucosidase reaction kinetics from isothermal titration microcalorimetry. Anal. Biochem. 2005, 347, 244–253.

51. Kristensen, J.B.; Felby, C.; Jørgensen, H. Yield-determining factors in high-solids enzymatic hydrolysis of lignocellulose. Biotechnol. Biofuels 2009, 2, 11.

52. Scheller, H.V.; Ulvskov, P. Hemicelluloses. Ann. Rev. Plant Biol. 2010, 61, 263–289.

53. Gao, D.; Uppugundla, N.; Chundawat, S.P.; Yu, X.; Hermanson, S.; Gowda, K.; Brumm, P.; Mead, D.; Balan, V.; Dale, B.E. Hemicellulases and auxiliary enzymes for improved conversion of lignocellulosic biomass to monosaccharides. Biotechnol. Biofuels 2010, 4, 5.

54. Couturier, M.; Haon, M.; Coutinho, P.M., Henrissat; Lesage-Meessen, L.; Berrin, J.-G. Podospora anserina hemicellulases potentiate the Trichoderma reesei secretome for saccharification of lignocellulosic biomass. Appl. Environ. Microbiol. 2011, 77, 237–246.

55. Gottschalk, L.M.F.; Oliveira, R.A.; Bon, E.P.D.S. Cellulases, xylanases, beta-glucosidase and ferulic acid esterase produced by Trichoderma and Aspergillus act synergistically in the hydrolysis of sugarcane bagasse. Biochem. Eng. J. 2010, 51, 72–78.

56. Kumar, R.; Wyman, C.E. Effect of xylanase supplementation of cellulase on digestion of corn stover solids prepared by leading pretreatment technologies. Bioresour. Technol. 2009, 100, 4203–4213.

57. Pollet, A.; Delcour, J.A.; Courtin, C.M. Structural determinants of the substrate specificities of xylanases from different glycoside hydrolase families. Crit. Rev. Biotechnol. 2010, 30, 176–191.

58. Ustinov, B.B.; Gusakov, A.V.; Antonov, A.I.; Sinitsyn, A.P. Comparison of properties and mode of action of six secreted xylanases from Chrysosporium lucknowense. Enzym. Microb. Technol. 2008, 43, 56–65.

59. Verjans, P.; Dornez, E.; Segers, M.; van Campenhout, S.; Bernaerts, K.; Beliën, T.; Delcour, J.A.; Courtin, C.M. Truncated derivatives of a multidomain thermophilic glycosyl hydrolase family 10 xylanase from Thermotoga maritima reveal structure related

activity profiles and substrate hydrolysis patterns. J. Biotechnol. 2010, 145, 160–167.

60. Jordan, D.B.; Wagschal, K. Properties and applications of microbial -D-xylosidases featuring the catalytically efficient enzyme from Selenomonas ruminantium. Appl. Microbiol. Biotechnol. 2010, 86, 1647–1658.

61. Pastor, F.I.J.; Gallardo, O.; Sanz-Aparicio, J.; Diaz, P. Xylanases: molecular properties and applications. In Industrial Enzymes; Polaina, J., MacCabe, A.P., Eds.; Springer: Dordrecht, the Netherlands, 2007; pp. 65–82.

62. Biely, P.; Mastihubova, M.; Tenkanen, M.; Eyzaguirre, J.; Li, X.L.; Vrsanska, M. Action of xylan deacetylating enzymes on monoacetyl derivatives of 4-nitrophenyl glycosides of -D-xylopyranose and -L-arabinofuranose. J. Biotechnol. 2011, 151, 137–142.

63. Koseki, T.; Fushinobu, S.; Ardiansyah, S.H.; Komai, M. Occurrence, properties, and applications of feruloyl esterases. Appl. Microbiol. Biotechnol. 2009, 84, 803–810.

64. Duranová, M.; Spániková, S.; Wösten, H.A.; Biely, P.; de Vries, R.P. Two glucuronoyl esterases of Phanerochaete chrysosporium. Arch. Microbiol. 2009, 191, 133–140.

65. Benoit, I.; Danchin, E.G.; Bleichrodt, R.J.; de Vries, R.P. Biotechnological applications and potential of fungal feruloyl esterases based on prevalence, classification and biochemical diversity. Biotechnol. Lett. 2008, 30, 387–396.

66. Saha, B.C. Alpha-L-arabinofuranosidases: biochemistry, molecular biology and application in biotechnology. Biotechnol. Adv. 2000, 18, 403–423.

67. Yip, V.L.Y.; Withers, S.G. Family 4 glycoside hydrolases are special: The first -elimination mechanism amongst glycoside hydrolases. Biocatal. Biotransform. 2006, 24, 167–176.

68. Chong, S.L.; Battaglia, E.; Coutinho, P.M.; Henrissat, B.; Tenkanen, M.; de Vries, R.P. The alpha-glucuronidase Agu1 from Schizophyllum commune is a member of a novel glycoside hydrolase family (GH115). Appl. Microbiol. Biotechnol. 2011, 90, 1323–1332.

69. Kolenova, K.; Ryabova, O.; Vrsanska, M.; Biely, P. Inverting

character of family GH115 -glucuronidases. FEBS Lett. 2010, 584, 4063–4068.

70. Martin, K.; McDougall, B.M.; McIlroy, S.; Chen, J.Z.; Seviour, R.J. Biochemistry and molecular biology of exocellular fungal beta-(1,3)- and beta-(1,6)-glucanases. FEMS Microbiol. Rev. 2007, 31, 168–192.

71. Moreira, L.; Filho, E. An overview of mannan structure and mannan-degrading enzyme systems. Appl. Microbiol. Biotechnol. 2008, 79, 165–178.

72. Baumann, M.J. Structural evidence for the evolution of xyloglucanase activity from xyloglucan endo-transglycosylases: biological implications for cell wall metabolism. Plant Cell 2007, 19, 1947–1963.

73. Kaida, R.; Kaku, T.; Baba, K.; Oyadomari, M.; Watanabe, T.; Nishida, K.; Kanaya, T.; Shani, Z.; Shoseyov, O.; Hayashi, T. Loosening xyloglucan accelerates the enzymatic degradation of cellulose in wood. Mol. Plant 2009, 2, 904–909.

74. Lombard, V.; Bernard, T.; Rancurel, C.; Brumer, H.; Coutinho, P.M.; Henrissat, B. A hierarchical classification of polysaccharide lyases for glycogenomics. Biochem. J. 2010, 432, 437–444.

75. Payasi, A.; Sanwal, R.; Sanwal, G.G. Microbial pectate lyases: characterization and enzymological properties. World J. Microbiol. Biotechnol. 2009, 25, 1–4.

76. Lundell, T.K.; Makela, M.R.; Hilden, K. Lignin-modifying enzymes in filamentous basidiomycetes—ecological, functional and phylogenetic review. J. Basic Microbiol. 2010, 50, 5–20.

77. Wymelenberg, A.V.; Gaskell, J.; Mozuch, M.; Sabat, G.; Ralph, J.; Skyba, O.; Mansfield, S.D.; Blanchette, R.A.; Martinez, D.; Grigoriev, I.; Kersten, P.J.; Cullen, D. Comparative transcriptome and secretome analysis of wood decay fungi Postia placenta and Phanerochaete chrysosporium. Appl. Environ. Microbiol. 2010, 76, 3599–3610.

78. Zamocky, M.; Ludwig, R.; Peterbauer, C.; Hallberg, B.M.; Divne, C.; Nicholls, P.; Haltrich, D. Cellobiose dehydrogenase—a flavocytochrome from wood-degrading, phytopathogenic and saprotropic fungi. Curr. Protein Pept. Sci. 2006, 7, 255–280.

79.    Bey, M.; Berrin, J.; Poidevin, L.; Sigoillot, J.-C. Heterologous expression of Pycnoporus cinnabarinus cellobiose dehydrogenase in Pichia pastoris and involvement in saccharification processes. Microb. Cell Factories 2011, 10, 113.

80.    Sweeney, M.D.; Vlasenko, E.; Abbate, E. Methods for increasing hydrolysis of cellulosic material in the presence of cellobiose dehydrogenase. U.S. Patent 20,100,159,536 A1, 24 June 2010.

81.    Langston, J.A.; Shaghasi, T.; Abbate, E.; Xu, F.; Vlasenko, E.; Sweeney, M.D. Oxidoreductive cellulose depolymerization by the enzymes cellobiose dehydrogenase and glycoside hydrolase 61. Appl. Environ. Microbiol. 2011, 77, 7007–7015.

82.    Phillips, C.; Beeson, W.; Cate, J.; Marletta, M. Cellobiose dehydrogenase and a copper dependent polysaccharide monooxygenase potentiate fungal cellulose. ACS Chem. Biol. 2011, 6, 1399–1406.

83.    Parawira, W.; Tekere, M. Biotechnological strategies to overcome inhibitors in lignocellulose hydrolysates for ethanol production: review. Crit. Rev. Biotechnol. 2011, 31, 20–31.

84.    Bendl, R.F.; Kandel, J.M.; Amodeo, K.D.; Ryder, A.M.; Woolridge, E.M. Characterization of the oxidative inactivation of xylanase by laccase and a redox mediator. Enzym. Microb. Technol. 2008, 43, 149–156.

85.    Xu, F.; Ding, H.; Tejirian, A. Detrimental effect of cellulose oxidation on cellulose hydrolysis by cellulase. Enzym. Microb. Technol. 2009, 45, 203–209.

86.    Saloheimo, M.; Nakari-Setala, T.; Tenkanen, M.; Penttila, M. cDNA cloning of a Trichoderma reesei cellulase and demonstration of endoglucanase activity by expression in yeast. Eur. J. Biochem. 1997, 249, 584–591.

87.    Karlsson, J.; Saloheimo, M.; Siika-Aho, M.; Tenkanen, M.; Penttila, M.; Tjerneld, F. Homologous expression and characterization of Cel61A (EG IV) of Trichoderma reesei. Eur. J. Biochem. 2001, 268, 6498–6507.

88.    Koseki, T.; Mese, Y.; Fushinobu, S.; Masaki, K.; Fujii, T.; Ito, K.; Shiono, Y.; Murayama, T.; Iefuji, H. Biochemical characterization of a glycoside hydrolase family 61 endoglucanase from Aspergillus kawachii. Appl. Microbiol. Biotechnol. 2008, 77, 1279–1285.

89.    Karkehabadi, S.; Hansson, H.; Kim, S.; Piens, K.; Mitchinson, C.;

Sandgren, M.L. The first structure of a glycoside hydrolase family 61 member, Cel61B from Hypocrea jecorina, at 1.6 A resolution. J. Mol. Biol. 2008, 383, 144–154.

90. Vaaje-Kolstad, G.; Houston, D.R.; Riemen, A.H.; Eijsink, V.G.; van Aalten, D.M. Crystal structure and binding properties of the Serratia marcescens chitin-binding protein CBP21. J. Biol. Chem. 2005, 280, 11313–11319.

91. Harris, P.V.; Welner, D.; McFarland, K.C.; Re, E.; Navarro Poulsen, J.C.; Brown, K.; Salbo, R.; Ding, H.; Vlasenko, E.; Merino, S.; Xu, F.; Cherry, J.; Larsen, S.; Lo Leggio, L. Stimulation of lignocellulosic biomass hydrolysis by proteins of glycoside hydrolase family 61: structure and function of a large, enigmatic family. Biochemistry 2010, 49, 3305–3316.

92. Sweeney, M.D.; Vlasenko, E. Methods for determining cellulolytic enhancing activity of a polypeptide. U.S. Patent 20,100,159,494, A1, 24 June 2010.

93. Forsberg, Z.; Vaaje-Kolstad, G.; Westereng, B.; Bunæs, A.C.; Stenstrøm, Y.; MacKenzie, A.; Sørlie, M.; Horn, S.J.; Eijsink, V.G. Cleavage of cellulose by a CBM33 protein. Protein Sci. 2011, 20, 1479–1483.

94. Quinlan, R.J.; Sweeney, M.D.; Lo Leggio, L.; Otten, H.; Poulsen, J.C.; Johansen, K.S.; Krogh, K.B.; Jørgensen, C.I.; Tovborg, M.; Anthonsen, A.; et al. Insights into the oxidative degradation of cellulose by a copper metalloenzyme that exploits biomass components. Proc. Natl. Acad. Sci. USA 2011, 108, 15079–15084.

95. Westereng, B.; Ishida, T.; Vaaje-Kolstad, G.; Wu, M.; Eijsink, V.G.; Igarashi, K.; Samejima, M.; Ståhlberg, J.; Horn, S.J.; Sandgren, M. The Putative Endoglucanase PcGH61D from Phanerochaete chrysosporium is a metal-dependent oxidative enzyme that cleaves cellulose. PLoS One 2011, 6, e27807.

96. Vaaje-Kolstad, G.; Westereng, B.; Horn, S.J.; Liu, Z.; Zhai, H.; Sørlie, M.; Eijsink, V.G. An oxidative enzyme boosting the enzymatic conversion of recalcitrant polysaccharides. Science 2010, 330, 219–222.

97. Sampedro, J.; Cosgrove, D.J. The expansin superfamily. Genome Biol. 2005, 6, 242.

98. Saloheimo, M.; Paloheimo, M.; Hakola, S.; Pere, J.; Swanson, B.; Nyyssönen, E.; Bhatia, A.; Ward, M.; Penttilä, M. Swollenin, a Trichoderma reesei protein with sequence similarity to the plant expansins, exhibits disruption activity on cellulosic materials. Eur. J. Biochem. 2002, 269, 4202–4211.

99. Wang, M.; Cai, J.; Huang, L.; Lv, Z.; Zhang, Y.; Xu, Z. High-level expression and efficient purification of bioactive swollenin in Aspergillus oryzae. Appl. Biochem. Biotechnol. 2010, 162, 2027–2036.

100. Jäger, G.; Girfoglio, M.; Dollo, F.; Rinaldi, R.; Bongard, H.; Commandeur, U.; Fischer, R.; Spiess, A.C.; Büchs, J. How recombinant swollenin from Kluyveromyces lactis affects cellulosic substrates and accelerates their hydrolysis. Biotechnol. Biofuels 2011, 23, 33.

101. Chen, X.A.; Ishida, N.; Todaka, N.; Nakamura, R.; Maruyama, J.; Takahashi, H.; Kitamoto, K. Promotion of efficient saccharification of crystalline cellulose by Aspergillus fumigatus Swo1. Appl. Environ. Microbiol. 2010, 76, 2556–2561.

102. Quiroz-Castañeda, R.E.; Martínez-Anaya, C.; Cuervo-Soto, L.I.; Segovia, L.; Folch-Mallol, J.L. Loosenin, a novel protein with cellulose-disrupting activity from Bjerkandera adusta. Microb. Cell Factories 2011, 10, 8.

103. Kerff, F.; Amoroso, A.; Herman, R.; Sauvage, E.; Petrella, S.; Filée, P.; Charlier, P.; Joris, B.; Tabuchi, A.; Nikolaidis, N.; Cosgrove, D.J. Crystal structure and activity of Bacillus subtilis YoaJ (EXLX1), a bacterial expansin that promotes root colonization. Proc. Natl. Acad. Sci. USA 2008, 105, 16876–16881.

104. Lee, H.J.; Lee, S.; Ko, H.J.; Kim, K.H.; Choi, I.G. An expansin-like protein from Hahella chejuensis binds cellulose and enhances cellulase activity. Mol. Cells 2010, 29, 379–385.

105. Kim, E.S.; Lee, H.J.; Bang, W.G.; Choi, I.G.; Kim, K.H. Functional characterization of a bacterial expansin from Bacillus subtilis for enhanced enzymatic hydrolysis of cellulose. Biotechnol. Bioeng. 2009, 102, 1342–1353.

106. Georgelis, N.; Tabuchi, A.; Nikolaidis, N.; Cosgrove, D.J. Structure-function analysis of the bacterial expansin EXLX1. J. Biol. Chem. 2011, 286, 16814–16823.

107. Foreman, P.K.; Brown, D.; Dankmeyer, L.; Dean, R.; Diener, S.; Dunn-Coleman, N.S.; Goedegebuur, F.; Houfek, T.D.; England, G.J.; Kelley, A.S.; et al. Transcriptional regulation of biomass-degrading enzymes in the filamentous fungus Trichoderma reesei. J. Biol. Chem. 2003, 278, 31988–31997.

108. Foreman, P.; van Solingen, P.; Goedegebuur, F.; Ward, M. CIP1 polypeptides and their uses. U.S. Patent 7,923,235, 12 April 2011. [Google Scholar]

109. Scott, B.R.; Hill, C.; Tomashek, J.; Liu, C. Enzymatic hydrolysis of lignocellulosic feedstocks using accessory enzymes. U.S. Patent 8,017,361 B2, 13 September 2011.

110. Li, X.L.; Špániková, S.; de Vries, R.P.; Biely, P. Identification of genes encoding microbial glucuronoyl esterases. FEBS Lett. 2007, 581, 4029–4035.

111. Fontes, C.M.; Gilbert, H.J. Cellulosomes: highly efficient nanomachines designed to deconstruct plant cell wall complex carbohydrates. Ann. Rev. Biochem. 2010, 79, 655–681.

112. Moraïs, S.; Barak, Y.; Hadar, Y.; Wilson, D.B.; Shoham, Y.; Lamed, R.; Bayer, E.A. Assembly of xylanases into designer cellulosomes promotes efficient hydrolysis of the xylan component of a natural recalcitrant cellulosic substrate. MBio 2011, 2, e00233–11.

113. Tsai, S.L.; Goyal, G.; Chen, W. Surface display of a functional minicellulosome by intracellular complementation using a synthetic yeast consortium and its application to cellulose hydrolysis and ethanol production. Appl. Environ. Microbiol. 2010, 76, 7514–7520.

114. Lilly, M.; Fierobe, H.P.; van Zyl, W.H.; Volschenk, H. Heterologous expression of a Clostridium minicellulosome in Saccharomyces cerevisiae. FEMS Yeast Res. 2009, 9, 1236–1249.

115. Anderson, T.D.; Robson, S.A.; Jiang, X.W.; Malmirchegini, G.R.; Fierobe, H.P.; Lazazzera, B.A.; Clubb, R.T. Assembly of minicellulosomes on the surface of Bacillus subtilis. Appl. Environ. Microbiol. 2011, 77, 4849–4858. [Google Scholar]

116. Zheng, Y.; Pan, Z.; Zhang, R. Overview of biomass pretreatment for cellulosic ethanol production. Int. J. Agric. Biol. Eng. 2009, 2, 51–68.

117. Zhang, T.; Datta, S.; Eichler, J.; Ivanova, N.; Axen, S.D.; Kerfeld,

C.A.; Chen, F.; Kyrpides, N.; Hugenholtz, P.; Cheng, J.F.; Sale, K.L.; Simmons, B.; Rubin, E. Identification of a haloalkaliphilic and thermostable cellulase with improved ionic liquid tolerance. Green Chem. 2011, 13, 2083–2090.

118. Graham, J.E.; Clark, M.E.; Nadler, D.C.; Huffer, S.; Chokhawala, H.A.; Rowland, S.E.; Blanch, H.W.; Clark, D.S.; Robb, F.T. Identification and characterization of a multidomain hyperthermophilic cellulase from an archaeal enrichment. Nat. Commun. 2011, 2, 375.

# Treatment Efficiency by Means of a Nonthermal Plasma Combined with Heterogeneous Catalysis of Odoriferous Volatile Organic Compounds Emissions from the Thermal Drying of Landfill Leachates

Matt D. Sweeney and Feng Xu

Daniel Almarcha,[1] Manuel Almarcha,[1] Elena Jimenez-Coloma,[2] Laura Vidal,[2] Montserrat Puigcercós[1], and Iban Barrutiabengoa[1]

[1]Ambiente y Tecnología Consultores, C. Còrsega 112, 08029 Barcelona, Spain

[2]Ferrovial, Avenida Catedral 6-8, 08002 Barcelona, Spain

# ABSTRACT

The objective of the present work was to assess the odoriferous volatile organic compounds depuration efficiency of an experimental nonthermal plasma coupled to a catalytic system used for odor abatement of real emissions from a leachate thermal drying plant installed in an urban solid waste landfill. VOC screening was performed by means of HRGC-MS analysis of samples taken at the inlet and at the outlet of the nonthermal plasma system. Odor concentration by means of dynamic olfactometry, total organic carbon, mercaptans, $NH_3$, and $H_2S$ were also determined in order to assess the performance of the system throughout several days. Three plasma frequencies (100, 150, and 200 Hz) and two catalyst temperatures (150°C and 50°C) were also tested. Under conditions of maximum capacity of the treatment system, the results show VOC depuration efficiencies around 69%, with average depuration efficiencies between 44 and 95% depending on the chemical family of the substance. Compounds belonging to the following families have been detected in the samples: organic acids, alcohols, ketones, aldehydes, pyrazines, and reduced sulphur compounds, among others. Average total organic carbon removal efficiency was 88%, while $NH_3$ and $H_2S$ removal efficiencies were 88% and 87%, respectively, and odor concentration abatement was 78%.

# INTRODUCTION

When rain water percolates through the solid waste of a landfill there is a process of dissolution and transport of water soluble elements and organic and inorganic substances, such as heavy metals, ammonium, inorganic anions, and volatile and semivolatile organic compounds, constituting a current that may end up reaching and significatively affecting the aquifers located in the peripheral areas of the facility and, therefore, potentially create risks to the environment and to human and animal health. A detailed summary of the composition of the leachates of urban solid waste (USW) landfills is included in [1], with data from the LEACH 2000 USA database.

Different approaches have been developed for the leachate treatment in order to mitigate the possible impact of USW landfills [2–

4]. The methods that are most often implemented (whether alone or in combination) are in-landfill recycling in order to increase the moisture of the waste that is being deposited and combined treatment with domestic sewage, aerobic and/or anaerobic biodegradation, chemical oxidation (including advanced oxidation), adsorption, precipitation, air stripping, microfiltration, ultrafiltration, nanofiltration, and reverse osmosis.

As an alternative to the mentioned techniques, in some cases thermal drying has also been used. It is based on an atomization process, during which $\cong 50\,\mu m$ droplets are formed, which are subjected to the action of a hot air current which causes the evaporation of the aqueous phase of the leachate from the surface of the droplets, thereby turning them into dry particles that, after a cooldown stage, are separated with a cyclone and a bag filter in series. As a residual effluent, a current of humid air with a "roasted" smell is produced. It has a high content of gaseous compounds with odoriferous significance, such as ammonia, amines, pyrazines, hydrogen sulfide, mercaptans, thioethers, carbonyls, alcohols, esters, and terpenes.

Preliminary studies concerning the characterization of the thermal drying process emissions of 3 urban solid waste (USW) leachate types allowed the determination of emission levels for several compounds with odoriferous significance, $H_2S$ (0.5–24 mg/Nm³), $NH_3$ (5–277 mg/Nm³), Alcohols $C_2$–$C_8$ (35–335 mg/Nm³), and Carbonyls $C_3$–$C_8$ (60–345 µg/Nm³), and emission levels of Pyrazines, which would explain the "roasted" odor descriptor assigned to the emission, between 5 and 55 µg/Nm³ [5]. Furthermore, odor concentration was determined by means of dynamic olfactometry as per the norm EN-13725:2005 and oscillated between 380 and 9,713 ou$_E$/m³ (with an average of 5,430 ou$_E$/m³), which, in the case of relatively high flow rates (as in the case of thermal drying of leachates, with flow rates >15,000 Nm³/h), means there is a possibility that there may be a considerable odor impact and, therefore, high efficacy emission treatment systems will have to be installed in those cases where the landfill surroundings are vulnerable. Given the composition and temperature of the emissions, some landfills have chosen high-performance odor depuration systems (with guaranteed efficiencies >95%) such as regenerative thermal oxidation (RTO) working with the landfill's own biogas. The main drawback associated with this technology, for this type of application, comes from the possible presence of significant levels of siloxanes [6] that, when burned inside the RTO system, are converted to $SiO_2$, which clogs the channels of the honeycomb ceramic heat exchangers. In order to avoid this problem, the exploration of new odor treatment alternatives must be considered. One alternative is nonthermal plasma (NTP) [7–13] with the possibility of coupling to supplementary treatments or posttreatments such as low-temperature catalytic oxidation and/or UV photooxidation.

Plasma is a gas which is partially or fully ionized and is made of particles such as electrons, atoms, ions, radicals, and molecules. While most components of thermal plasma are in a state of thermal equilibrium, in the case of nonthermal plasmas, the plasma temperature used for quantitatively describing the plasma has not reached said thermal equilibrium and presents substantial differences in its content of electrons and other particles. NTPs may be generated by means of several methods, such as dielectric barrier discharge (DBD), corona discharge, pulsed corona, and microwaves. The electrons are accelerated by means of an electric field at typical temperatures between 10,000 and 250,000 K, and, under these conditions, they induce the formation of free radicals from parent molecules, which in turn react with the effluent pollutants and decompose them. Chemical reactions involving free radicals typically last less than $10^{-3}$ s, which can translate to highly compact systems. Normal pollutant concentrations are around the 100 s of ppm, and therefore the direct interaction between the electrons and the pollutant molecules is negligible. The NTP induces chemical reactions at moderate conditions of approximately 1 atm and room temperature. The addition of oxidant substances is not required because the NTPs use the oxygen and water vapor to generate the reactive radicals.

In order to implement applications with significant yields, systems have been developed which combine the NTP with the use of adsorbents (such as activated carbon), different catalysts, and also supplementary UV postoxidation. The combination of a nonthermal plasma with catalytic processes (NTP-CAT) can be carried out by (a) in-plasma catalysis (IPC), where the selected catalyst is put directly inside the discharge chamber or by (b) postplasma catalysis (PPC), placing the catalyst after the discharge chamber. A heterogeneous catalyst is used, including $Al_2O_3$, $SiO_2$, $TiO_2$, $MnO_2$, platinum-based catalysts, and modified zeolites, and it can be introduced in the reactor in different forms. The details of the theoretical and practical aspects of NTP and NTP-CAT have been abundantly discussed in the available bibliography.

Table 1 includes a summary of the publications cited in the review article by Van Durme et al. [14] on the maximum abatement efficiencies for several VOC treated by means of coupled dielectric barrier discharge (DBD) nonthermal plasmas, both IPC and PPC, with different catalysts (DBD-NTP-CAT). It must be taken into account that most of the data in Table 1 and other publications cited in this work comes from laboratory tests treating currents with a single target pollutant [14–23].

**Table 1**: VOC abatement efficiencies with DBD-NTP-CAT tests

| Compounds | Catalyst | Maximum abatement efficiency % | Publications cited by Van Durme et al. (2008) in [14]* |
|---|---|---|---|
| Benzene | $TiO_2$ | 60 | Chae et al., 2004 (28) |
| | $TiO_2/SiO_2$ | 50 | |
| | $TiO_2$ | 16 | Lu et al., 2006 (4) |
| | $Ag/TiO_2$ | 96 | Kim et al., 2005 (12) |
| Toluene | $MnO_2/Al_2O_3$ | 55 | Delagrange et al., 2006 (26) |
| | $CuO/MnO_2$ | 50 | Chae et al., 2004 (28) |
| | $TiO_2$ | 16 | Lu et al., 2006 (4) |
| | $MnO_x / Co_x$ | >99 | Subrahmanyan et al., 2006 (2) |
| $SF_6$ | $CuO/ZnO/MgO/Al_2O_3$ | >99 | Chang et al., 2004 (29) |
| $NF_3$ | | >99 | |
| $CF_4$ | | 66 | |
| $C_2F_6$ | | 83 | |
| HCOH | $Ag/CeO_2$ | 92 | Ding et al., 2006 (30) |
| Trichloroethene (TCE) | Au-SBA-15 | >99 | Magureanu et al., 2007 (34) |
| TCE | $MnO_2$ | 97 | Subrahmanyan et al., 2007 (2) |
| Isopropanol | $MnO_x / CoO_x$ | >99 | Han et al., 2007 (35) |
| Dichloromethane | Various | 34–51 | Intriago et al., 2007 (36) |

*The bracketed italic numbers correspond to the reference numbers in Van Durme et al. (2008) [14], and not the references of this paper.

In the present work a DBD-NTP-CAT pilot plant was installed in the Palautordera landfill (Barcelona, Spain), where the urban solid waste leachate is treated by means of thermal drying. The gas effluents emitted from the drying process are treated by means of an RTO system that uses biogas as fuel. A diagram of the installation is shown in Figure 1, including the two configurations that were tested (indicated as Options A and B, resp.). As will be explained in the following section, it must be noted that the quantitative efficiency tests were carried out with the Option B configuration.

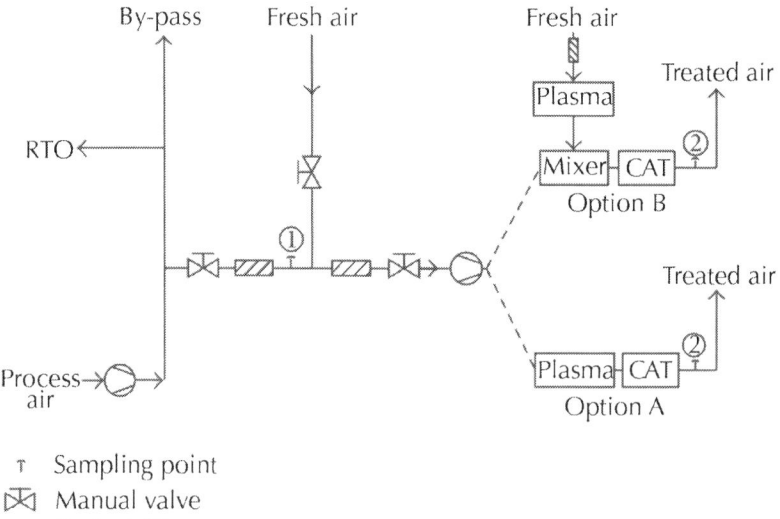

**Figure 1**: Diagram of flow process of the NTP-CAT pilot plant configurations tested.

# PLASMA OPERATION CONDITIONS

Preliminary tests were carried out with the configuration indicated as Option A in Figure 1, that is, conducting the emission (diluted with clean air) directly into the plasma chamber, from where it was transferred to the chamber containing the catalyst. With this mode of operation it was found that, when the gas entered into contact with the electrodes, abundant small water droplets would condense that, because of the high operation electrical potentials, generated sparks which damaged the electrodes and made it impossible to carry out the corresponding efficacy tests. It was decided, therefore, that only ambient air would enter the plasma chamber. Then, the ionized ambient air was led into a mixing chamber together with the emission air as indicated in Figure 1 (Option B configuration). This was the configuration that was finally adopted, and it must be pointed out that this strategy requires an increase in both the size of the equipment and the electric consumption, meaning higher CAPEX and OPEX costs. The effluent gas flow rate coming from the thermal drying system was $100 \, m^3/h$ (with a residence time <2 s). $MnO_2$ was used as catalyst and was tested at two different fixed temperatures: 50°C and 150°C. The global dilution rate was between 1:5 and 3:5, v:v.

Four campaigns were carried out using the Option B configuration described above. The first 3 campaigns (designated in Table 2 as Day 1, Day 2, and Day 3, resp.) were testing the conditions which were foreseen to provide the highest depuration efficiencies (200 Hz plasma frequency, 1:5 (v:v) dilution ratio, and 150°C catalyst temperature). The fourth campaign (Day 4 in Table 2) intended to provide data on the performance of the system under conditions that minimized the electrical cost associated to the heating of the catalyst (3:5 (v:v) dilution ratio and 50°C catalyst temperature). This fourth campaign was also meant to assess the effect of a decrease of the plasma frequency.

**Table 2:** Odour concentration, $VOC_{OD}$, $NH_3$, $H_2S$, and TOC concentrations, and % reduction in each day

| Parameter | Day 1 | Day 2 | Day 3 | Average (Days 1–3) | Day 4.1 | Day 4.2 | Day 4.3 | Day 4.4 |
|---|---|---|---|---|---|---|---|---|
| Odour conc. (dynamic olfactom.) | | | | | | | | |
| Inlet conc. ($ou_E/m^3$) | 13,777 | 38,968 | 20,632 | 24,459 | — | — | — | — |
| Outlet conc. ($ou_E/m^3$) | 5,161 | 2,580 | 4,812 | 4,184 | — | — | — | — |
| Reduction (%) | 63 | 93 | 77 | 78 | — | — | — | — |
| $\sum VOC_{OD}$ | | | | | | | | |
| Inlet conc. ($\mu g/m^3$) | 148.4 | 470.6 | 195.9 | 271.6 | 98.3 | 92.6 | 41.2 | 42.5 |
| Outlet conc. ($\mu g/m^3$) | 51.8 | 145.8 | 55.2 | 84.3 | 45.8 | — | 19.3 | — |
| Average $VOC_{OD}$ concentration reduction (%) | 65 | 69 | 72 | 69 | 53 | — | 53 | — |
| $NH_3$ | | | | | | | | |
| Inlet conc. (ppmV) | 139.3 | 143 | 115 | 132.4 | 265.2 | 265.2 | 170.5 | 170.5 |
| Outlet conc. (ppmV) | 18.1 | 19.2 | 13 | 16.8 | 50.5 | 40.6 | 60.2 | 43.7 |
| Reduction (%) | 87 | 87 | 89 | 87 | 81 | 85 | 65 | 74 |
| $H_2S$ | | | | | | | | |
| Inlet conc. (ppmV) | 3.7 | 4.0 | 2.0 | 3.2 | <0.5 | <0.5 | <0.5 | <0.5 |
| Outlet conc. (ppmV) | <0.5 | <0.5 | <0.5 | <0.5 | <0.5 | <0.5 | <0.5 | <0.5 |
| Reduction (%) | >86 | >88 | >75 | >83 | n.a.* | n.a.* | n.a.* | n.a.* |

| TOC | | | | | | | | |
|---|---|---|---|---|---|---|---|---|
| Inlet conc. (mgC/m³) | 62.3 | 46.5 | 51.5 | 53.4 | 167.3 | 150.1 | 140 | 140 |
| Outlet conc. (mgC/m³) | 10.5 | 5.7 | 3.5 | 6.6 | 72.3 | 129.3 | 102.5 | 120.8 |
| Reduction (%) | 83 | 88 | 93 | 88 | 57 | 14 | 27 | 14 |
| Plasma frequency (Hz) | 200 | 200 | 200 | — | 200 | 150 | 200 | 100 |
| Effluent:dilution air ratio (v:v) | 1:5 | | | | 3:5 | | | |
| Catalyst (temperature) | MnO₂ (150°C) | | | | MnO₂ (50°C) | | | |

*n.a.: not applicable; the compound was not detected at the inlet or at the outlet.
Note: the results indicated as < (value) correspond to the LOD of the corresponding technique.

# MATERIALS AND METHODS FOR THE DETERMINATION OF THE TREATMENT EFFICIENCY OF THE NTP-CAT PILOT PLANT

## Sampling and "In Situ" Measurements

The samples were taken at the sampling points 1 and 2, as shown in Figure 1. Sampling point 1 was located at the inlet of the system and corresponds to the process gas before the dilution with fresh air, while sampling point 2 was located at the outlet of the system after the treatment. As the residence time of the gas in the system was very short, the inlet and outlet samples were taken simultaneously.

On-site determinations of hydrogen sulphide ($H_2S$), mercaptans (RSH), and ammonia ($NH_3$) were carried out by means of "ad hoc" colorimetric tubes (Dräger) by pumping the required sample volumes through the tubes. Limits of detection were 0.5 ppmV for RSH, 0.2 ppmV for $H_2S$, and 2 ppmV for $NH_3$. In order to determine the regularity of the gaseous effluent to be treated and the final emission, the total organic carbon (TOC) was also measured "in situ" with a portable flame ionization detector (FID) analyzer (Thermo FID, Mess-Analysentechnik Gmbh).

Two fractions were considered regarding the VOC studied. The first fraction includes the VOC with a molecular size between 2 and 6 carbon atoms ($C_2$–$C_6$), which was sampled by pumping 15 L of gas through an activated charcoal cartridge (ORBO, 250 mg). The fraction containing the VOC with more than 6 carbon atoms () was sampled by filling 5 L Nalophan bags by means of indirect aspiration with a lung-like device. All the samples were transported to the laboratory and analyzed as quickly as possible.

The 10 L samples contained in Nalophan bags for dynamic olfactometry analysis were taken using the same procedure described in the preceding paragraph for the VOC > $C_6$ and were sent immediately to the laboratory in a refrigerated and temperature-controlled container so they could be analyzed within 30 h, as indicated in the norm EN-13725:2005 [24].

# Sample Treatment and Analysis

## *VOC Screening*

HRGC-MS analyses were performed by means of a ThermoTrace HRGC-MS system with a 30 m × 0.25 mm ID × 1.4 μm BPX-624 column (SGE). Chromatographic conditions for both the analysis of the activated charcoal and the samples contained in Nalophan bags were carrier gas He at 1.5 mL/min, injection temperature 245°C, temperature program 35°C (5 min), and 5°C/min up to 250°C, staying at 250°C for 10 minutes. MS acquisition was performed in Full-Scan mode (m/z range 35 to 350 a.m.u.). Data treatment and identification of the analytes were carried out using the Xcalibur software (Thermo Fisher) with NIST and Wiley spectra libraries. The identification of an analyte was only considered positive if there was at least 85% matching between its spectrum and the library one.

Each of the VOC fractions was analyzed with the following conditions.

(1)   *Activated Charcoal Samples*: The charcoal where the VOC were adsorbed is transferred to a vial and desorbed with 1,2,3-trimethylbenzene (Sigma Aldrich, p.a.) containing Toluene-D8 as internal standard. The quantitation limit was 0.1 μg/m³.

(2)   *Samples Contained in Nalophan Bags*: First, the samples are brought to the laboratory, where a Toluene-D8 internal standard (Supelco) is added to a concentration of 25 μg/m³. Then a PDMS fiber (Supelco) was exposed to the sample for 60 minutes in order for the analytes to be adsorbed onto the fiber, after which the fiber was inserted into the HRGC system's injector at 245°C for the desorption of the adsorbed VOCs to take place. The quantitation limit was 0.1 μg/m³.

(3)   *Quantitation*: The quantitation of the analytes present in both the activated charcoal and the Nalophan bag samples was carried out by using standards of some 100 compounds belonging to

the VOC families that are commonly found in environmental samples (including alcohols, esters, carbonyls, carboxylic acids, aromatic compounds, terpenes, and thioethers). These standards were used for the calculation of calibration curves for the corresponding compounds and relative response factors between each compound and Toluene-D8. Quantitation was then carried out by relative response to the Toluene-D8 internal standard, using relative response factors for most analytes. As it was impossible to have standards for all of the compounds that are typically present in this kind of effluents (and it is also impossible to know beforehand which ones will be present), the relative response factors of similar compounds were used when analytes were detected for which no standard was available.

## Odor Concentration Analyses

Odor concentration was determined in an accredited lab by means of Dynamic Olfactometry according to the requirements indicated in the norm EN-13725:2005 [24].

Details of the applied techniques can be found in [25–35].

## Results

A summary of the results of the VOC analyses, as well as the other studied parameters, is included in this section.

The VOC results shown in Tables 2 and 3 do not include all of the compounds of this group that were detected, and only the compounds with low enough odour thresholds to present significant odor implications, according to our experience [36], have been included in the study. These relevant odoriferous VOC ($VOC_{OD}$) belong for the most part to the following families: pyrazines and pyridines, organic acids, alcohols and ethers, aldehydes, ketones, esters, and reduced sulphur compounds (mercaptans and thioethers). Other nonodoriferous compounds were detected in the samples, such as aliphatic and aromatic hydrocarbons, but have not been taken into account, as they have higher odour thresholds [37] and do not significantly contribute to an emission's odor impact [5, 27].

**Table 3**: Inlet and outlet concentration ranges and average concentration reduction for each compound family

| Family | Inlet conc. range (µg/m³) | Outlet conc. range (µg/m³) | Average conc. reduction (%) |
|---|---|---|---|
| Organic acids* | 3.7–29.5 | 0.2–3.5 | 67.7 |
| Alcohols and ethers | 32.7–227.7 | 6.9–33.3 | 68.0 |
| Aldehydes | 2.1–232.8 | 0.5–5.5 | 65.3 |
| Ketones | 20.5–143.5 | 2.7–34.6 | 69.0 |
| Esters | 0.4–2.1 | 0.1–1.0 | 44.7 |
| Pyrazines and pyridines | 2.2–54.3 | 0.2–24.1 | 54.2 |
| Sulphur compounds | 1.3–5.5 | n.d. <0.1 | 95.3 |
| Other | 1.0–10.6 | 0.2–2.2 | 64.5 |

*Excluding acetic acid, that had higher outlet than inlet levels, possibly due to new formation.

The experimental work was carried out during four days. During the first three days, the NTP system worked all the time with a plasma frequency of 200 Hz, and all of the indicated parameters were analyzed. During the fourth day, three frequencies were tested (100, 150, and 200 Hz, resp.). It must be pointed out that speciated $VOC_{OD}$ analyses were only performed at 200 Hz, while TOC, $NH_3$, $H_2S$, and mercaptans were determined at each of the three studied frequencies.

Table 2 shows the $VOC_{OD}$ concentration reduction, which corresponds to the average of all the individual $VOC_{OD}$ reductions. Also shown are the plasma frequencies corresponding to each assay, as well as the % reductions of TOC, $NH_3$, and $H_2S$ in each day. The concentrations of mercaptans determined "in situ" were always under the detection limit (0.5 ppmV) in all of the samples analyzed during this study and therefore were not included in the table, but it must be pointed out that mercaptans were not detected in any of the HRGC-MS analyses of the charcoal cartridges, with LOD around 2 orders of magnitude smaller. Finally, a summary of the plasma operation conditions is also included in the table (catalyst type, temperature, and effluent/clean air dilution rates). Table 3 shows the $VOC_{OD}$ families present in the samples, along with the mean reduction efficacy for each kind of compound. Generally, the most relevant families in terms of concentrations were alcohols, aldehydes, ketones, and organic acids, and there was also a highly significant presence of pyrazines, which are usually formed in thermal drying processes such as the one carried out in the plant where the NTP system was tested and which have a strong roasted-like odor [5]. Their inlet concentrations were between 0.1 and 40.1 µg/m$^3$. It must be pointed out that, for certain compounds, their concentration in the outlet samples may sometimes be higher than in the inlet or even newly appear and therefore individually have negative % reductions.

Figure 2 shows two speciated VOC$_{OD}$ () TIC chromatograms corresponding to the SPME analyses carried out at the inlet and the outlet of the NTP system samples taken during Day 3.

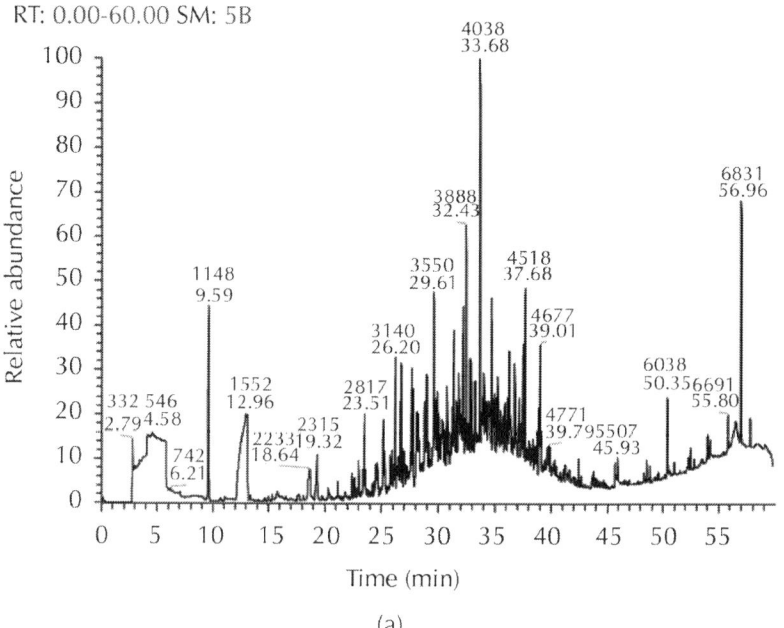

(a)

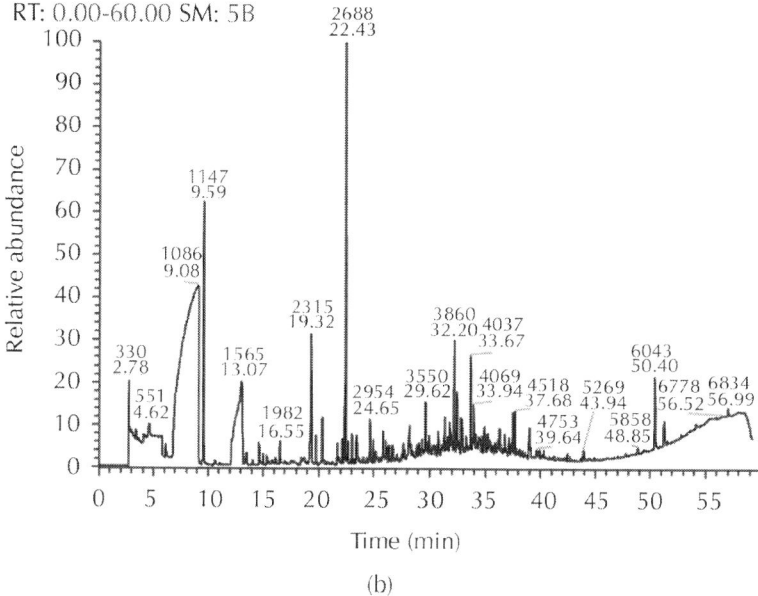

RT: 0.00-60.00 SM: 5B

(b)

**Figure 2**: TIC chromatograms of an inlet (a) and outlet (b) sample.

As an example, Table 4 shows the speciated $VOC_{OD}$ () screening results corresponding to Day 4.1, along with the % reduction for each compound.

**Table 4**: Odoriferous $VOC_{OD}$ $(C_n>C_6)$ screening results and % reduction, corresponding to Day 4.1

| Compound | Inlet conc. | Outlet conc. | % |
|---|---|---|---|
| | µg/m³ | µg/m³ | Reduction |
| 2-Hexanone | 0.2 | <0.1 | >50.0 |
| 2-Hexen-1-ol acetate | 0.4 | 0.4 | 0.0 |
| 2-Hexen-1-ol | 1.7 | 1.2 | 29.4 |
| Dimethylpyridine | 0.6 | 0.8 | −33.3 |
| Cyclohexanone | 1.3 | <0.1 | >92.3 |
| 2-Ethylpyridine | 0.1 | <0.1 | 0.0 |
| 2,6-Dimethylpyrazine | 0.9 | 0.6 | 33.3 |
| 2,3-Dimethylpyrazine | 0.9 | 0.6 | 33.3 |
| 2-Ethyl-2-hexenal | 0.2 | 0.2 | 0.0 |

| Benzaldehyde | 0.2 | <0.1 | >50.0 |
|---|---|---|---|
| 2-Ethylhexanal | 0.5 | <0.1 | >80.0 |
| Hexanoic acid | 0.3 | 0.2 | 33.3 |
| Terpene | 0.3 | 0.2 | 33.3 |
| 2,4,6-Trimethylpyridine | 0.3 | <0.1 | >66.7 |
| 2-Methyl-6-ethylpyrazine | 0.4 | 0.2 | 50.0 |
| Trimethylpyrazine | 13.7 | 8.9 | 35.0 |
| Limonene | 0.8 | <0.1 | >87.5 |
| Eucalyptol | 1.3 | 0.8 | 38.5 |
| Dihydroisophorone | 0.6 | <0.1 | >83.3 |
| 2-Ethyl-3,5-dimethylpyrazine | 4.9 | 3 | 38.8 |
| 3-Ethyl-2,5-dimethylpyrazine | 19.5 | 9.8 | 49.7 |
| Alcohol | 2.4 | 1.9 | 20.8 |
| 2-Ethylhexanoic acid | 3.5 | <0.1 | >97.1 |
| Camphor | 2.7 | 1.7 | 37.0 |
| Octanoic acid | 3.5 | 3.4 | 2.9 |
| 3-Undecanone | 6.9 | 4.8 | 30.4 |
| Possible sulphur derivative | 5.5 | <0.1 | >98.2 |
| 2-Methyl-1-dodecanol | 4.8 | 3.1 | 35.4 |
| Alcohol | 14.3 | 8.7 | 39.2 |
| $(\sum VOC_{OD}(C_n > C_6))$ | 92.7 | 51.5 | 44.4 |

# DISCUSSION AND CONCLUSIONS

As shown in Table 2, the $NH_3$ depuration efficiency, working in the conditions of maximum capacity of the treatment system, was quite good, ≥87%, which is similar to those reported in [10] of 93.5%, in [38] of 80%, and in [36] of ≅100%. The results shown in [39], however, are significantly different, with a reported $NH_3$ depuration efficiency of 44%. The average $H_2S$ reduction that was determined during this work is >83%, which is comparable to the results found in [8, 39], which are 96.3% and 91%, respectively.

As can be seen in Table 3, the % reductions for the $VOC_{OD}$ concentrations were in the 45–95% range, while Table 2 shows that the average total % reductions are between 53% and 75%. These depuration efficiencies may prove to be somewhat low when dealing with critical emissions or those with high odor loads. These values are close to those reported in [40], where depuration efficiencies for thioethers were 78–100% and 32–51% for other VOC and with those in [39] which indicate abatement efficiencies in the 45–93% range for oxygenated VOC such as aldehydes, ketones, and alcohols. The average total organic carbon content reduction of 88% shows good agreement with the determined VOC reduction efficiencies achieved during this study, although it must be pointed out that these results are significantly different from the 35% TOC reduction reported in [39].

Regarding the odor concentration abatement, the results published in [40] (70–90%) and [39] (72–88%) are similar to our results (78%). Finally, the dynamic olfactometry and $VOC_{OD}$ data shows that the reductions are quantitatively comparable (with reduction values of 78% and 89%, resp.).

The results obtained in different plasma conditions during Day 4 show that a lower catalyst temperature, together with a higher effluent: dilution air ratio, translates to a small decrease in the $NH_3$ abatement efficacy, as well as a more significant decrease in the $VOC_{OD}$ and TOC reductions. Furthermore, the experiment carried out during Day 4 shows that lower plasma frequencies correspond to a clear decrease in the TOC reduction efficacy.

As has been previously mentioned, the most important $VOC_{OD}$ families detected in the samples, in terms of concentration, were alcohols, aldehydes, organic acids, and pyrazines. The highest depuration efficiencies were those corresponding to sulphur compounds and ketones (95% and 69% on average, resp.), while esters and pyrazines (the latter compounds have an olfactive descriptor coincident with that of the studied emission) showed the poorest removals with the NTP-CAT system studied during the present work, with values just near 50% in the experimental conditions that were used. In this regard, it must be pointed out that these reduction percentages show good agreement with those published in different references (e.g., [38, 40]).

On the other hand, even though it falls outside the scope of the present work, an aspect that must be highlighted is the fact that the levels of depuration efficiencies of odoriferous compounds do not correspond with the experimentally observed odor concentration reductions determined in the laboratory by means of dynamic olfactometry (carried out according to the norm EN-13725:2003) as, while a clear reduction of odor intensity between the samples taken before and after the NTP-CAT system could be organoleptically perceived, this was not clearly reflected by the results of the olfactometric assays. The causes that may explain these not fully satisfactory odor depuration efficiencies may be attributed, among others, to a possible incomplete degradation of some $VOC_{OD}$ and that some of the resulting by-products may have lower olfactive thresholds (OTV) than their precursors (e.g., [39, 41]), which may be due to the selected operation configuration and/or the catalyst that was chosen for this pilot test. It must be pointed out that, in some cases, a new formation of oxidized species (such as acetic acid or some carbonyls) was observed, but, due to the limited number of samples that were analyzed, it was not possible to arrive at definitive conclusions, and these compounds were excluded from the $VOC_{OD}$ lists.

It can be concluded, therefore, that the NTP-CAT technology that was tested, in its current state and under the applied conditions, will likely not be capable of providing sufficient performance in critical deodorization operations (e.g., high odor or $VOC_{OD}$ loads, very close populated receptors, etc.) which require very stringent demands regarding odor depuration efficiency (even as high as >95%). For instance, for highly loaded emissions (generally >20,000 $uo_E/m^3$), it would be impossible to achieve final emissions with odor concentrations higher than those typically requested in environmental permits (usually around 1,000 $uo_E/m^3$). However, it may be an interesting technology in less demanding situations or as an addition, as a previous stage or backup, to other odor treatment systems [39, 42]. It may also be useful for the treatment of emissions which contain particles or aerosols with odor implications, as the nonthermal plasma treatment has been shown to remove a significant part of the particulate matter from the emission [43, 44], which may have adsorbed odoriferous compounds, as in the cases of composting plants or WWTP sludge drying facilities. There are also publications that state that the treatment by means of an NTP also eliminates the microorganisms associated to the bioaerosols [38, 45].

# ACKNOWLEDGMENTS

The authors acknowledge the Spanish Ministerio de Medio Ambiente for the financial support of the Project with reference 099/PC08/2-02.2 of the Programa Nacional de Proyectos de Desarrollo Experimental en el marco del Plan Nacional de Investigación Científica, Desarrollo e Innovación Tecnológica, 2008–2011.

# REFERENCES

1.  Electric Power Research Institute (EPRI), "Comparison of risks for leachate from coal combustion product landfills and impoundments with risks for leachate from municipal solid waste landfill facilities," Final Report, 2010.

2.  A. A. Abbas, G. Jingsong, L. Z. Ping, P. Y. Ya, and W. S. Al-Rekabi, "Review on landfill leachate treatments," The American Journal of Applied Sciences, vol. 6, no. 4, pp. 672–684, 2009.

3.  S. Renou, J. G. Givaudan, S. Poulain, F. Dirassouyan, and P. Moulin, "Landfill leachate treatment: review and opportunity," Journal of Hazardous Materials, vol. 150, no. 3, pp. 468–493, 2008.

4.  Environment Agency, EHS (Northern Ireland), and The Scottish Environment Protection Agency (SEPA), "Guidance for the treatment of landfill leachate," Sector Guidance Note IPPC S5.03, 2007.

5.  D. Almarcha, E. Jimenez-Coloma, and J. Caixach, "Determination of odoriferous compounds in the emissions from a termal dry treatment of leachates in urban solid waste landfill," in Proceedings of the 3rd International Symposium on Energy from Biomass and Waste, Venecia, Colo, USA, Noviembre 2010.

6.  R. Dewil, L. Appels, and J. Baeyens, "Energy use of biogas hampered by the presence of siloxanes," Energy Conversion and Management, vol. 47, no. 13-14, pp. 1711–1722, 2006.

7.  J. S. Chang, "Recent development of plasma pollution control technology: a critical review," Science and Technology of Advanced Materials, vol. 2, no. 3-4, pp. 571–576, 2001.

8.   X. Feng and C. Jie-rong, "Application of non-thermal plasma technology for indoor air pollution control," Environmental Informatics Archives, vol. 2, pp. 628–634, 2004.

9.   H.-H. Kim, "Nonthermal plasma processing for air-pollution control: a historical review, current issues, and future prospects," Plasma Processes and Polymers, vol. 1, no. 2, pp. 91–110, 2004.

10.  U.S. Environmental Protection Agency, "Using non-thermal plasma to control air pollutants," Tech. Rep. EPA-456/R-05-001, U.S. Environmental Protection Agency, 2005.

11.  J. A. Beukes, "Applied plasma physics and atmospheric chemical principles in odor removal," Chemical Engineering Transactions, vol. 15, pp. 419–421, 2008.

12.  Y.-H. Bai, J.-R. Chen, X.-Y. X.-Y. Li, and C.-H. Zhang, "Non-thermal plasmas chemistry as a tool for environmental pollutants abatement," Reviews of Environmental Contamination and Toxicology, vol. 201, pp. 117–136, 2009.

13.  A. M. Vandenbroucke, R. Morent, N. de Geyter, and C. Leys, "Non-thermal plasmas for non-catalytic and catalytic VOC abatement," Journal of Hazardous Materials, vol. 195, pp. 30–54, 2011.

14.  J. Van Durme, J. Dewulf, C. Leys, and H. van Langenhove, "Combining non-thermal plasma with heterogeneous catalysis in waste gas treatment: a review," Applied Catalysis B: Environmental, vol. 78, no. 3-4, pp. 324–333, 2008.

15.  M. P. Cal and M. Schluep, "Destruction of benzene with non-thermal plasma in dielectric barrier discharge reactors," Environmental Progress, vol. 20, no. 3, pp. 151–156, 2001.

16.  S. Pekárek, "Non-thermal plasma ozone generation," Acta Polytechnica, vol. 43, no. 6, pp. 47–51, 2003.

17.  T. Zhu, J. Li, Y. Jin, Y. Liang, and G. Ma, "Decomposition of benzene by non-thermal plasma processing: photocatalyst and ozone effect," International Journal of Environmental Science and Technology, vol. 5, no. 3, pp. 375–384, 2008.

18.  M. Magureanu, N. B. Mandache, V. I. Parvulescu, C. Subrahmanyam, A. Renken, and L. Kiwi-Minsker, "Improved performance of non-thermal plasma reactor during decomposition of trichloroethylene: optimization of the reactor geometry

and introduction of catalytic electrode," Applied Catalysis B: Environmental, vol. 74, no. 3-4, pp. 270–277, 2007.

19. C. Subrahmanyam, M. Magureanu, A. Renken, and L. Kiwi-Minsker, "Catalytic abatement of volatile organic compounds assisted by non-thermal plasma. Part 1. A novel dielectric barrier discharge reactor containing catalytic electrode," Applied Catalysis B: Environmental, vol. 65, no. 1-2, pp. 150–156, 2006.

20. C. Subrahmanyam, A. Renken, and L. Kiwi-Minsker, "Catalytic abatement of volatile organic compounds assisted by non-thermal plasma. Part II. Optimized catalytic electrode and operating conditions," Applied Catalysis B: Environmental, vol. 65, no. 1-2, pp. 157–162, 2006.

21. W. Mista and R. Kacprzyk, "Decomposition of toluene using non-thermal plasma reactor at room temperature," Catalysis Today, vol. 137, no. 2-4, pp. 345–349, 2008.

22. J. Karuppiah, E. L. Reddy, P. M. K. Reddy, B. Ramaraju, and C. Subrahmanyam, "Catalytic nonthermal plasma reactor for the abatement of low concentrations of benzene," International Journal of Environmental Science and Technology, vol. 11, no. 2, pp. 311–318, 2014.

23. B. Dou, F. Bin, C. Wang, Q. Jia, and J. Li, "Discharge characteristics and abatement of volatile organic compounds using plasma reactor packed with ceramic Raschig rings," Journal of Electrostatics, vol. 71, no. 5, pp. 939–944, 2013.

24. "Air quality: determination of odour concentration by dynamic olfactometry," Tech. Rep. EN-13725:2003, 2003.

25. E. Davoli, M. L. Gangai, L. Morselli, and D. Tonelli, "Characterisation of odorants emissions from landfills by SPME and GC/MS," Chemosphere, vol. 51, no. 5, pp. 357–368, 2003.

26. C. Easter, C. Quigley, P. Burrowes, J. Witherspoon, and D. Apgar, "Odor and air emissions control using biotechnology for both collection and wastewater treatment systems," Chemical Engineering Journal, vol. 113, no. 2-3, pp. 93–104, 2005.

27. D. Almarcha, M. Almarcha, S. Nadal, and J. Caixach, "Comparison of the depuration efficiency for voc and other odoriferous compounds in conventional and advanced biofilters in the abatement of odour emissions from municipal waste treatment

plants," Chemical Engineering Transactions, vol. 30, pp. 259–264, 2012.

28. J. A. Koziel, L. Cai, D. W. Wright, and S. J. Hoff, "Solid-phase microextraction as a novel air sampling technology for improved, GC-olfactometry-based assessment of livestock odors," Journal of Chromatographic Science, vol. 44, no. 7, pp. 451–457, 2006.

29. B. Scaglia, V. Orzi, A. Artola, et al., "Odours and volatile organic compounds emitted from municipal solid waste at different stage of decomposition and relationship with biological stability," Bioresource Technology, vol. 102, no. 7, pp. 4638–4645, 2011.

30. A. Kännaste, L. Copolovici, and Ü. Niinemets, "Gas chromatography-mass spectrometry method for determination of biogenic volatile organic compounds emitted by plants," Methods in Molecular Biology, vol. 1153, pp. 161–169, 2014.

31. A. Ribes, G. Carrera, E. Gallego, X. Roca, M. J. Berenguer, and X. Guardino, "Development and validation of a method for air-quality and nuisance odors monitoring of volatile organic compounds using multi-sorbent adsorption and gas chromatography/mass spectrometry thermal desorption system," Journal of Chromatography A, vol. 1140, no. 1-2, pp. 44–55, 2007.

32. Ö. O. Kuntasal, D. Karman, D. Wang, S. G. Tuncel, and G. Tuncel, "Determination of volatile organic compounds in different microenvironments by multibed adsorption and short-path thermal desorption followed by gas chromatographic-mass spectrometric analysis," Journal of Chromatography A, vol. 1099, no. 1-2, pp. 43–54, 2005.

33. K. Demeestere, J. Dewulf, K. de Roo, P. de Wispelaere, and H. van Langenhove, "Quality control in quantification of volatile organic compounds analysed by thermal desorption-gas chromatography-mass spectrometry," Journal of Chromatography A, vol. 1186, no. 1-2, pp. 348–357, 2008.

34. P. Ciccioli, E. Brancaleoni, M. Frattoni, A. Cecinato, and L. Pinciarelli, "Determination of volatile organic compounds (VOC) emitted from biomass burning of Mediterranean vegetation species by GC-MS," Analytical Letters, vol. 34, no. 6, pp. 937–955, 2001.

35. C. Rodríguez-Navas, R. Forteza, and V. Cerdà, "Use of thermal desorption-gas chromatography-mass spectrometry (TD-GC-MS) on identification of odorant emission focus by volatile organic compounds characterisation," Chemosphere, vol. 89, no. 11, pp. 1426–1436, 2012.

36. J.-J. Ruan, W. Li, Y. Shi, Y. Nie, X. Wang, and T.-E. Tan, "Decomposition of simulated odors in municipal wastewater treatment plants by a wire-plate pulse corona reactor," Chemosphere, vol. 59, no. 3, pp. 327–333, 2005.

37. S. Murnane, A. H. Lehocky, and P. D. Owens, Odor Thresholds for Chemicals with Established Occupational Health Standards, American Industrial Hygiene Association, 2nd edition, 2013.

38. C. W. Park, J. H. Byeon, K. Y. Yoon, J. H. Park, and J. Hwang, "Simultaneous removal of odors, airborne particles, and bioaerosols in a municipal composting facility by dielectric barrier discharge," Separation and Purification Technology, vol. 77, no. 1, pp. 87–93, 2011.

39. M. Hołub, R. Brandenburg, H. Grosch, S. Weinmann, and B. Hansel, "Plasma supported odour removal from waste air in water treatment plants: an industrial case study," Aerosol and Air Quality Research, vol. 14, no. 3, pp. 697–707, 2014.

40. K. B. Andersen, A. Feilberg, and J. A. Beukes, "Use of non-thermal plasma and UV-light for removal of odour from sludge treatment," Water Science and Technology, vol. 66, no. 8, pp. 1656–1662, 2012.

41. K. B. Andersen, A. Feilberg, and J. A. Beukes, "Abating odor nuisance from pig production units by the use on a non-thermal plasma system," Chemical Engineering Transactions, vol. 23, pp. 351–356, 2010.

42. H. Kim, B. Han, W. Hong, J. Ryu, and Y. Kim, "A new combination system using biotrickling filtration and nonthermal plasma for the treatment of volatile organic compounds," Environmental Engineering Science, vol. 26, no. 8, pp. 1289–1297, 2009.

43. J.-S. Chang, "Next generation integrated electrostatic gas cleaning systems," Journal of Electrostatics, vol. 57, no. 3-4, pp. 273–291, 2003.

44. J. H. Byeon, J. Hwang, J. Hong Park et al., "Collection of submicron

particles by an electrostatic precipitator using a dielectric barrier discharge," Journal of Aerosol Science, vol. 37, no. 11, pp. 1618–1628, 2006.

45.   R. Haumacher and W. P. ReinhardBöhm, "Bioaerosols from composting-quantitative measurements on biofilters and non-thermal plasma technology," in Proceedings of the 12th ISHA Congress, vol. 2, pp. 276–278, Warsaw, Poland, 2005.

# Citations

# CHAPTER 1

Adam F. Lee, James A. Bennett, Jinesh C. Manayil, and Karen Wilson, Heterogeneous catalysis for sustainable biodiesel production via esterification and transesterification, DOI: 10.1039/C4CS00189C.

# CHAPTER 2

Andrzej Wieckowski and Matthew Neurock, Contrast and Synergy between Electrocatalysis and Heterogeneous Catalysis, Advances in Physical Chemistry, vol. 2011, Article ID 907129, 18 pages, 2011. doi:10.1155/2011/907129.

# CHAPTER 3

Le Tu Thanh, Kenji Okitsu, Luu Van Boi, and Yasuaki Maeda, Catalytic Technologies for Biodiesel Fuel Production and Utilization of Glycerol: A Review, doi:10.3390/catal2010191

# CHAPTER 4

Yanyong Liu , Rogelio Sotelo-Boyás , Kazuhisa Murata , Tomoaki Minowa and Kinya Sakanishi, Production of Bio-Hydrogenated Diesel by Hydrotreatment of High-Acid-Value Waste Cooking Oil over Ruthenium Catalyst Supported on Al-Polyoxocation-Pillared, doi:10.3390/catal2010171.

# CHAPTER 5

René Bindig, Saad Butt, Ingo Hartmann, Mirjam Matthes, and Christian Thiel, Application of Heterogeneous Catalysis in Small-Scale Biomass Combustion Systems, doi:10.3390/catal2020223.

# CHAPTER 6

Matt D. Sweeney and Feng Xu, Biomass Converting Enzymes as Industrial Biocatalysts for Fuels and Chemicals: Recent Developments, doi: 10.3390/catal2020244.

# CHAPTER 7

Daniel Almarcha, Manuel Almarcha, Elena Jimenez-Coloma, Laura Vidal, Montserrat Puigcercós, and Iban Barrutiabengoa, Treatment Efficiency by means of a Nonthermal Plasma Combined with Heterogeneous Catalysis of Odoriferous Volatile Organic Compounds Emissions from the Thermal Drying of Landfill Leachates, Journal of Engineering, vol. 2014, Article ID 831584, 9 pages, 2014. doi:10.1155/2014/831584.

# Index